T0250542

Sanitation of the Harvesting, Processing and Distribution of Shellfish

National Shellfish Sanitation Program Manual of Operations, Part II

Revised 1992

■

Center for Food Safety and Applied Nutrition,
Shellfish Sanitation Branch

United States Food and Drug Administration

The Interstate Shellfish Sanitation Conference

C. K. SMOLEY

Library of Congress Cataloging-in-Publication Data

Center for Food Safety and Applied Nutrition (U.S.). Shellfish Sanitation Branch.

National Shellfish Sanitation Program manual of operations / Center for Food Safety and Applied Nutrition Shellfish Sanitation Branch, United States Food and Drug Administration, the Interstate Shellfish Sanitation Conference. — Rev. 1992.

v. <1–2>

Includes bibliographical references.

Contents: pt. 1. Sanitation of shellfish growing areas — pt. 2. Sanitation of the harvesting, processing, and distribution of shellfish.

ISBN 0-8493-8724-8 (pt. 1) — ISBN 0-8493-8725-6 (pt. 2)

1. National Shellfish Sanitation Program (U.S.) — Handbooks, manuals, etc. 2. Shellfish fisheries — United States — Sanitation — Standards. I. United States. Food and Drug Administration. II. Interstate Shellfish Sanitation Conference. III. Title.
RA602.S6C45 1993
363.19'29—dc20

93-16122
CIP

© 1993 by C. K. SMOLEY

Direct all inquiries to CRC Press, Inc., 2000 Corporate Blvd., N.W., Boca Raton, Florida 33431.

PRINTED IN THE UNITED STATES OF AMERICA
2 3 4 5 6 7 8 9 0
Printed on acid-free paper

LIST OF PREVIOUS EDITIONS OF MANUAL OF OPERATIONS FOR NATIONAL SHELLFISH SANITATION PROGRAM - NOW SUPERSEDED

1925.   Supplement No. 53 to Public Health Reports November 6, 1925 "Report of Committee on Sanitary Control of Shellfish Industry in the United States".

1937.   U.S. Public Health Service Minimum Requirement for Approval of State Shellfish Control Measures and Certification for Shippers in Interstate Commerce (Revised October 1937).

1946.   Manual of Recommended Practice for Sanitary Control of the Shellfish Industry Recommended by the U.S. Public Health Service (Public Health Bulletin No. 295).

1957.   Manual of Recommended Practice for Sanitary Control of the Shellfish Industry (Part II: Sanitation of the Harvesting and Processing of Shellfish). Printed as Part II of Public Health Service Publication No. 33.

1959.   Manual of Recommended Practice for Sanitary Control of the Shellfish Industry (Part I: Sanitation of Shellfish Growing Areas). Printed as Part I of Public Health Service Publication No. 33.

1962.   Cooperative Program for the Certification of Interstate Shellfish Shippers, Part II, Sanitation of the Harvesting and Processing of Shellfish. (Printed as Part II of Public Health Service Publication No. 33).

1962.   Cooperative Program for the Certification of Interstate Shellfish Shippers, Part I, Sanitation of Shellfish Growing Areas. (Printed as Part I of Public Health Service Publication No. 33).

1965.   National Shellfish Sanitation Program Manual of Operations Part I, Sanitation of Shellfish Growing Areas, Public Health Service Publication No. 33, revised 1965.

1965.   National Shellfish Sanitation Program Manual of Operations Part II, Sanitation of the Harvesting and Processing of Shellfish, Public Health Service Publication No. 33, revised 1965.

1965.   National Shellfish Sanitation Program Manual of Operations Part III, Public Health Service Appraisal of State Shellfish Sanitation Programs, Public Health Service Publication No. 33, revised 1965.

1986.   National Shellfish Sanitation Program Manual of Operations Part I, Sanitation of Shellfish Growing Areas, revised 1986.

1987.   National Shellfish Sanitation Program Manual of Operations Part II, Sanitation of the Harvesting, Processing, and Distribution of Shellfish, revised 1987.

1988.   National Shellfish Sanitation Program Manual of Operations Part I, Sanitation of Shellfish Growing Areas, revised 1988.

1988.   National Shellfish Sanitation Program Manual of Operations Part II, Sanitation of the Harvesting, Processing, and Distribution of Shellfish, revised 1988.

1989.   National Shellfish Sanitation Program Manual of Operations Part I, Sanitation of Shellfish Growing Areas, revised 1989.

1989.   National Shellfish Sanitation Program Manual of Operations Part II, Sanitation of the Harvesting, Processing, and Distribution of Shellfish, revised 1989.

1990.   National Shellfish Sanitation Program Manual of Operations Part I, Sanitation of Shellfish Growing Areas, revised 1990.

1990.   National Shellfish Sanitation Program Manual of Operations Part II, Sanitation of the Harvesting, Processing, and Distribution of Shellfish, revised 1990.

(This page is blank.)

This is Part II of the two volume National Shellfish Sanitation Program *Manual of Operations* published by the Food and Drug Administration

Part I - *Sanitation of Shellfish Growing Areas* Revised 1992

Part II - *Sanitation of the Harvesting, Processing and Distribution of Shellfish* Revised 1992

(This page is blank.)

CONTENTS

(This page is blank.)

CONTENTS (cont.)

(This page is blank.)

CONTENTS (cont.)

(This page is blank.)

FOREWORD

A Declaration of Principles

Oysters, clams and mussels are unique foods which have been enjoyed by consumers for many years. The popularity of shellfish as a food can be traced through several centuries of American history. To early settlers, the food resources of the sea were one of the most valuable and readily usable of the natural resources, particularly from the estuaries. It is not surprising that shellfish were foremost among their staple food items.

The value of these renewable natural resources to the early settlers was reflected in colonial legislation designed to encourage their wise use. Over 300 years ago in 1658, the Dutch Council of New Amsterdam passed an ordinance regulating the taking of oysters from the East River. Other early legislation, including that of New York (1715), New Jersey (1730), and Rhode Island (1734), was designed to regulate harvesting, presumably as conservation measures to guarantee a continuing supply.

Public health controls of shellfish became a national concern in the U.S. in the late 19th and early 20th century when public health authorities noted a large number of illnesses associated with consuming raw oysters, clams, and mussels. These shellfish-associated outbreaks were also medically recorded in other parts of the world, most notably in European countries. During the winter of 1924, there occurred a widespread typhoid fever outbreak, with cases in New York, Chicago, and Washington, D.C., which was finally traced to sewage polluted oysters. Local and State public health officials, and the shellfish industry became sufficiently alarmed over this outbreak to request that the Surgeon General of the United States Public Health Service develop necessary control measures to ensure a safe shellfish supply to the consuming public.

In accordance with this request, the Surgeon General called a conference of representatives from State and municipal health authorities, State conservation commissions, the Bureau of Chemistry (later to become the Food and Drug Administration), the Bureau of Commercial Fisheries (now National Marine Fisheries Service) and the shellfish industry. This historic conference was held in Washington, D.C. on February 19, 1925.

The members of the conference recommended eight resolutions for the sanitary control of the oyster industry. These included:

> "The beds on which shellfish are grown must be determined, inspected, and controlled by some official State agency and the U.S. Public Health Service."

> "The plants in which shellfish are shucked or otherwise prepared or packed by the shipper must be inspected and controlled by some official State agency and the U.S. Public Health Service."

"There must be such governmental supervision and such trade organization as will make plain the source of shellfish and will prevent shellfish from one source being substituted for those from another source. This will be chiefly a problem of the individual State."

"The methods of shipping must be supervised, inspected, controlled and approved by the proper official federal and State agency."

"The product must conform to an established bacterial standard and must meet federal, State, and local laws and regulations relative to salinity, water content, food proportion and conform to the Pure Food Laws standards."

The conference also established a committee to develop further necessary guidelines to recommend control practices for the sanitary control of the shellfish industry.

The basic concepts in formulating a program of national public health controls were reiterated by the Surgeon General in his letter of August 12, 1925, to State health officers and all others concerned. This letter set forth the following understandings:

1. "The Public Health Service considers that the responsibility for the sanitary control of the shellfish industry rests chiefly upon the individual States; and that the requisite coordination and uniformity of control may best be achieved by mutual agreement among the States, with the assistance and cooperation of the Public Health Service..."

2. "In accordance with this principle, it is considered that each producing State is directly responsible for the effective regulation of all production and handling of shellfish within its confines, not merely for the protection of its own citizens, but equally for safeguarding such of its product as goes to other States..."

3. "In order that each State may have full information concerning the measures carried out in other States, the Public Health Service will undertake systematic surveys of the machinery and efficiency of sanitary control as actually established in each producing State, and will report thereon for the information of the authorities of other States. It is believed that, in addition to furnishing valuable information, these reports will have an important influence in stimulating the development of better sanitary control and in promoting substantial uniformity on a higher plane."

2. "The officers of the Public Health Service assigned to this survey work will assist the State agencies in determining their sanitary problems, in formulating plans for adequate sanitary control, and in making actual sanitary surveys as far as practicable."

4. "In addition to the above, the Public Health Service will continue to extend the services which

it is already rendering, especially in conducting scientific investigations of fundamental importance to control, and in serving as a clearinghouse for the interchange of information and the discussion of policies between State authorities."

To implement this program, the members of the 1925 conference agreed that the producing States would issue "Certificates," i.e., a permit to operate, to shellfish shippers that meet agreed upon sanitary standards. The Public Health Service would serve as a clearinghouse for information on the effectiveness of the State control programs. This clearinghouse responsibility was met initially through issuance of a periodic "Progress Report on Shellfish Sanitation" describing the shellfish sanitation program in each State. This procedure was subsequently abandoned in favor of a "program endorsement" concept. Under this concept, the Public Health Service made a continuing appraisal of each State's shellfish sanitation program to determine if the control measures were in substantial accord with the provisions of the current "Manual of Recommended Practice for Sanitary Control of the Shellfish Industry." The Public Health Service also published a list of all shellfish shippers certified by those States that maintained "satisfactory" control programs.

The procedures used by the Public Health Service in fulfillment of its obligations under the Public Heath Service Act resulted from an understanding that implementation and enforcement of the necessary public health controls could best be accomplished under State laws with federal technical support and industry participation. The National Shellfish Sanitation Program is dependent entirely upon the States adopting the recommended requirements and the cooperative and voluntary efforts of State regulatory agencies and the shellfish industry.

(This page is blank.)

INTRODUCTION

The National Shellfish Sanitation Program (NSSP) developed from public health principles and Program controls formulated at the original conference on shellfish sanitation called by the Surgeon General of the United States Public Health Service in 1925. These fundamental components were described in a supplement to *Public Health Reports*, "Report of Committee on Sanitary Control of the Shellfish Industry in the United States" (1925)(1).

The public health control procedures established by the Public Health Service were dependent on the cooperative and voluntary efforts of State regulatory agencies. These efforts were augmented by the assistance and advice of the Public Health Service (now the Food and Drug Administration) and the voluntary participation of the shellfish industry. These three parties combined to form a tripartite cooperative program.

To carry out this cooperative control program, each partner accepted responsibility for certain procedures.

> Each shellfish shipping State adopted adequate laws and regulations for sanitary control of the shellfish industry, completed sanitary surveys of growing areas, delineated and patrolled restricted areas, inspected shellfish plants, and conducted such additional inspections, laboratory investigations, and control measures as were necessary to insure that the shellfish reaching the consumer had been grown, harvested and processed in a sanitary manner. The State annually issued numbered certificates to shellfish dealers who complied with the agreed-upon sanitary standards, and forwarded copies of the interstate certificates to the FDA.

> The FDA made an annual review of each State shellfish control program including the inspection of a representative number of shellfish processing plants. On the basis of the information thus obtained, the FDA determined the degree of conformity the State control program had with the NSSP. For the information of health authorities and others concerned, the FDA published a monthly list of valid interstate shellfish shipper certificates.

> The shellfish industry cooperated by obtaining shellfish from safe sources, by providing plants which met the agreed upon sanitary standards, by maintaining sanitary operating conditions, by placing the proper certificate number on each package of shellfish, and by keeping and making available to the control authorities records which showed the origin and disposition of all shellfish.

Although the basic public health principles of the NSSP have remained unchanged, program procedures have been updated and improved upon at periodic intervals. The original 1925 "Report of Committee on Sanitary Control of the Shellfish Industry in the United States" was revised and reissued in 1937 and again in 1946. The document was then divided into two parts Part II entitled

"Sanitation of Harvesting and Processing of Shellfish" was issued in 1957 and in 1959, Part I, "Sanitation of Shellfish Growing Areas." The need for a specialized program of this nature was reaffirmed by the cooperating members at the First National Shellfish Sanitation Workshop held in Washington, D.C., in 1954 (2) and at subsequent National Shellfish Sanitation Workshops held in 1956 (3), 1958 (4), 1961 (5) and 1964 (6).

The 1965 revision of the shellfish sanitation manual was prepared in cooperation with the shellfish control authorities in all coastal States, food control authorities in the inland States, interested federal agencies, Canadian federal departments, the Oyster Institute of North America, the Pacific Coast Oyster Growers Association, and the Oyster Growers and Dealers Association of North America.

In 1968, the 6th National Shellfish Sanitation Workshop was held (7). Recommendations for further revisions to the 1965 Manual were made and accepted by Workshop participants. This Workshop was structured around 12 task forces that were assigned specific topics to examine and provide recommendations to the general assembly. This approach to examining and discussing a large number of issues resulted in many actions being considered by the NSSP. It was recommended that this approach be used for future Workshops.

The shellfish sanitation program responsibilities assigned to the Assistant Secretary for Health, Department of Health, Education and Welfare were delegated to the Commissioner of Food and Drugs in late 1968. The FDA continued to sponsor the National Shellfish Sanitation Workshops in 1971, 1974, 1975, and 1977 (8, 9, 10, 11). Proceedings from these Workshops contain additional recommendations for revision to the 1965 Manual of Operations.

On June 19, 1975, the FDA proposed National Shellfish Safety Program Regulations in the *Federal Register* (12). There was considerable discussion at the 1975 and 1977 Workshops concerning these proposed regulations, and after evaluation of the comments received as a result of the proposed rules, the FDA determined that promulgating federal regulations would not likely achieve NSSP goals. Subsequently, FDA decided to revise the 1965 Manual of Operations as a better approach to strengthening the NSSP. (See Federal Register of February 26, 1985, 50 F.R. 7797)

During this period, many shellfish producing States were concerned that some State shellfish control agencies were not adopting revisions and enforcing NSSP guidelines in a uniform and timely manner. These States and FDA began exploring methods for strengthening the NSSP by means other than adopting the proposed regulations. In reviewing other means, it was noted that since 1950, a successful voluntary public health program, the National Conference of Interstate Milk Shippers (NCIMS), has been operating to assure a nationwide safe and wholesome milk supply. This program received direction and advice from the NCIMS.

The success of the NCIMS program prompted State shellfish control officials and FDA to select the NCIMS program as a model for developing an organization. In 1982, a delegation of State officials from 22 States met in Annapolis, Maryland and formed the Interstate Shellfish Sanitation Conference (ISSC). The ISSC is composed of state shellfish regulatory officials, industry officials, FDA, and other federal agencies.

The ISSC organization permits State regulatory officials to establish uniform guidelines and to exchange reliable information on sources of safe shellfish. The first annual meeting was held in New Orleans, Louisiana in August, 1983. At this conference, the ISSC adopted the NSSP Manual, as well as formal procedures that will enable it to adopt changes in the Manual. In March, 1984, FDA entered into a Memorandum of Understanding (MOU) with the ISSC and formally established a cooperative relationship with both the States and shellfish industry. At the second annual meeting in Orlando, Florida in August, 1984, the Conference accepted for review a revision of Part I of the NSSP Manual of Operations. At the third annual meeting in Cherry Hill, New Jersey, in August, 1985, the Conference adopted an Update Part I of the NSSP Manual of Operations, and accepted for review a revision of Part II of the NSSP Manual. The ISSC will continue to play an important role in assuring that uniform shellfish control measures are adopted, and that those measures are enforced consistently by state regulatory authorities.

In revising Part II of the 1965 NSSP Manual of Operations, FDA relied principally on the following sources:

(1) The draft revision of the Proposed National Shellfish Safety Program Regulations, Part 951;

(2) The NSSP Update *Manual of Operations, Part I, Sanitation of Shellfish Growing Areas*, and the 1965 NSSP *Manual of Operations, Part II, Sanitation of the Harvesting and Processing of Shellfish*, and *Part III, Appraisal of State Shellfish Sanitation Programs*, U.S. Department of Health, Education, and Welfare, Public Health Service Publication No. 33;

(3) The National Shellfish Sanitation Program Workshop Proceedings for 1968, 1971, 1973, 1974, and 1977;

(4) The Environmental Protection Agency rules and regulations (40 CFR Parts 400, et seq.) concerning water pollution control and shellfish waters;

(5) Other federal laws and regulations concerning quality of shellfish and shellfish growing areas;

(6) Existing State rules and regulations concerning shellfish growing area control and water quality criteria; and

(7) Analytical methods accepted by the American Public Health Association, Association of Official Analytical Chemists, American Society of Testing Materials, and other voluntary standard-setting organizations relating to shellfish and shellfish waters.

In updating the 1965 Manual, it was recognized that the growing and processing of shellfish were two distinct phases of operation in the shellfish industry. Therefore, the current Manual continues to be prepared in two parts; Part I: *Sanitation of Shellfish-Growing Areas*; and Part II: *Sanitation of the Harvesting, Processing and Distribution of Shellfish.* Part I of the Manual is a guide for preparing State shellfish laws and regulations pertaining to sanitary control of shellfish growing area classification, laboratory procedures, relaying, patrol operations and marine biotoxins. Part II of the Manual is a guide for operating, inspecting and certifying shellfish shippers, processors and depuration facilities; and for controlling interstate shipments of shellfish.

States participating in the NSSP will continue to be guided by Parts I and II of the Update Manual for exercising supervision over shellfish growing and relaying areas, and in the issuing of certificates to shellfish shippers. The Update Manual will also be used by FDA in evaluating State shellfish sanitation programs to determine if the programs conform with the recommended guidelines of the NSSP.

Developing the Update Manual was a cooperative effort between FDA and the ISSC. Initial drafts were prepared by FDA and presented to the ISSC and other interested parties for review and comment. The comments were incorporated into drafts after consultation with the ISSC, and the final revision was presented to the ISSC for formal endorsement. Further modifications to the Update Manual will reflect mutual FDA and ISSC concurrence. It is envisioned that after Parts I and II of the Update Manual and a Model Shellfish Sanitation Code are adopted by FDA and the ISSC, that the NSSP and the Interstate Shellfish Sanitation Program (ISSP) will be merged into a single national program.

In addition to setting forth the principles and requirements for the sanitary control of shellfish produced and shipped in interstate commerce in the U.S., the Update Manual is intended to be used by the States to control the harvesting and handling of shellfish for recreational and intrastate commercial use. Most coastal States believe that consumers residing in their State should be provided equal public health protection as are consumers in other States under the interstate certification program. To accomplish this States apply the same water quality and harvesting restrictions on non-interstate shellfish activities as on interstate activities. Having uniform intra and interstate programs also greatly facilitates the effective implementation and regulation of all shellfish harvesting activities, and results in the most efficient utilization of public health resources.

The Update Manual is also intended to be used by FDA as the basis for "certifying" foreign shellfish sanitation programs. To accomplish this, FDA seeks to establish international MOUs with official agencies in those foreign countries that wish to export shellfish to the U.S. An MOU is established after the foreign government demonstrates to FDA that the government has laws or regulations equivalent to those published in the Manual, and that the foreign program is supported by trained personnel, laboratory facilities, and other resources as may be necessary to exercise control over the export shellfish industry. Once a country has an effective MOU, the shellfish control authority submits certificates of their certified shellfish dealers to the FDA. The FDA publishes the names of these certified shellfish shippers in the Interstate Certified Shellfish Shippers List as an approved source of shellfish.

Definitions

Adequate - Means that which is needed to accomplish the intended purpose in keeping with good public health practice.

Air Gap - The unobstructed vertical distance through the free atmosphere between the lowest opening from any pipe or faucet supplying water to a tank, plumbing fixture or other device and the flood level rim of the receptacle.

Approved Area - The classification of a state shellfish growing area which has been approved by the state shellfish control authority for growing or harvesting shellfish for direct marketing. The classification of an approved area is determined through a sanitary survey conducted by the state shellfish control authority in accordance with Part I Section C of this Manual. An approved classified growing area may be temporarily made a closed area when a public health emergency such as a hurricane or flooding is declared.

Aquaculture - The controlled production of molluscan shellfish in natural and artificial systems. Components of aquaculture may overlap with other activities covered in the Manual such as relaying, transplanting, wet storage, depuration, growing water classification and labeling.

Backflow - The flow of water or other liquids, mixtures, or substances into the distributing pipes of a potable supply of water from any source or sources other than its intended source.

Backsiphonage - The flowing back of used, contaminated, or polluted water from a plumbing fixture or vessel or other source into a potable water supply pipe due to negative pressure in such pipes.

Blower - A container for washing shucked shellfish which uses forced air as a means of agitation.

Certification - The issuing by the SSCA of a numbered license or permit to operate that indicates compliance with the sanitation and program requirements of the NSSP. Certification of a shipper assures receiving jurisdictions that a firm meets NSSP criteria and is therefore eligible for interstate shipment and listing in the ICSSL.

Certification Number - The number assigned by the state shellfish control agency to each certified shellfish dealer. It consists of a one to five digit number preceded by the two letter state abbreviation and followed by the two letter symbol designating the type of operation certified. The SSCA may issue a certification number to all firms with separate facilities based on meeting the standards set forth in the NSSP.

Closed Area - A growing area where the harvesting of shellfish is temporarily or permanently not permitted. A closed area status is or may be placed on any of the five classified area designations approved, conditionally approved, restricted, conditionally restricted, or prohibited.

Commingling - The act of combining different lots of shellfish or shucked shellfish.

Conditionally Approved Area - The classification of a state shellfish growing area determined by the state shellfish control authority to meet approved area criteria for a predictable period. The period is conditional upon established performance standards specified in a management plan. A conditionally approved shellfish growing area is a closed area when the area does not meet the approved growing area criteria and is temporarily closed by the shellfish control authority.

Conditionally Restricted Area - The classification of a state shellfish growing area determined by the state shellfish control authority to meet restricted area criteria for a predictable period. The period is conditional upon established performance standards specified in a management plan. A conditionally restricted shellfish growing area is a closed area when the area does not meet the restricted growing area criteria and is temporarily closed by the shellfish control authority.

Depuration - The process of using a controlled, aquatic environment to reduce the level of bacteria and viruses in live shellfish.

Corrosion-resistant Materials - Those materials that maintain their original surface characteristics under normal exposure to the foods being contacted, normal use of cleaning compounds and bactericidal solutions, and other conditions of use.

Critical deficiency - A condition or practice which: a) results in the production of a product which is unwholesome; *or* b) presents a threat to the health or safety of consumers.

Cross connection - Any physical connection or arrangement between two (2) otherwise separate piping systems, one of which contains potable water and the other, water of unknown or questionable safety, or steam, gas or chemical whereby there may be a flow from one system to the other, the direction of flow depending on the pressure differential between the two systems.

Dealer - A commercial shellfish shipper, reshipper, shucker-packer, repacker, or depuration processor or operation.

Depuration Plant - A depuration plant is a facility of one or more depuration units. A depuration unit is a tank or series of tanks supplied by a single process water system.

Depuration Processor (DP) - A person who receives shellstock from approved or restricted growing areas and submits such shellstock to an approved depuration process.

Dry Storage - The storage of shellstock out of water.

Easily Cleanable - A surface which is readily accessible; and is made of such materials, has a finish and is so fabricated that residues may be effectively removed by normal cleaning methods.

Food-contact surfaces - An equipment surface or utensil with which shucked shellfish normally come into contact, directly or indirectly.

Harvester - A person who takes shellfish by any means from a growing area.

Heat shock - The process of subjecting shellstock to any form of heat treatment, such as steam, hot water or dry heat for a short period of time prior to shucking to facilitate removal of the meat from the shell without substantially altering the physical or organoleptic characteristics of the shellfish.

Key deficiency - A condition or practice which may result in adulterated, decomposed, misbranded or unwholesome product.

Label - Any written, printed, or graphic matter affixed to or appearing upon any package containing shellfish.

License - The document issued by the appropriate State shellfish control agency which authorizes a person to harvest and transport shellfish for commercial sale.

Lot of Shellstock - A collection of bulk shellstock or containers of shellstock of no more than one day's harvest from a single defined growing area harvested by one or more harvesters.

Lot of Shellstock for Depuration - Shellstock harvested from a particular area at a particular time and delivered to one depuration plant.

Lot of Shucked Shellfish - A collection of containers of no more than one day's shucked shellfish product produced under conditions as nearly uniform as possible, and designated by a common container code or marking.

Marine Biotoxins - Poisonous compounds accumulated by shellfish feeding upon toxic microorganisms. The poisons may come from dinoflagellates, e.g. *Alexandrium spp.* (formerly *Protogonyaulax spp.*, *Gonyaulax catenella* and *G. tamarensis*) and *Gymnodinium breve* (formerly *Ptychodiscus brevis*).

National Shellfish Sanitation Program (NSSP) - The cooperative State-FDA-Industry program for certification of interstate shellfish shippers as described in the NSSP Manual of Operations, Part I and II. Foreign governments may be members by having a current MOU or agreement with the FDA.

Other deficiency - A condition or practice that is not in accordance with NSSP Manual requirements but is not key or critical.

Person - An individual, partnership, corporation, association or other legal entity.

Poisonous or Deleterious Substance - A toxic compound occurring naturally or added to the environment that may be found in shellfish and for which a regulatory tolerance limit has been or may be established to protect public health. Examples of naturally occurring substances would be shellfish toxins and trace elements such as mercury; and examples of added substances would be agricultural pesticides and polynuclear aromatics from oil spills.

Process Batch - A quantity of shellfish used to fill each separate depuration unit.

Process Water - The water in depuration tanks during the time that shellfish are being depurated.

Prohibited Area - State waters that have been classified by the state shellfish control agency as prohibited for the harvesting of shellfish for any purpose except depletion. A prohibited shellfish growing area is a closed area for harvesting shellfish at all times.

Principal Display Panel - The part of a label that is most likely to be displayed, presented, shown or examined under customary conditions for retail sale.

Processor - A person who depurates, shucks, packs, or repacks shellfish.

Repacker (RP) - A person other than the original certified shucker-packer who repacks shucked shellfish into other containers. A repacker may also repack and ship shellstock. A repacker shall not shuck shellfish.

Reshipper (RS) - A person who purchases shucked shellfish or shellstock from other certified shippers and sells the product without repacking or relabeling to other certified shippers, wholesalers, or retailers.

Restricted Area - State waters that have been classified through a sanitary survey by the state shellfish control agency as an area from which shellfish may be harvested only if permitted and subjected to a suitable and effective relay or depuration process.

Safe Materials - Articles manufactured from or composed of materials that may not reasonably be expected to result, directly or indirectly, in their becoming a component or otherwise affecting the characteristics of any food.

Sanitize - The treatment to adequately treat food-contact surfaces by a process that is effective in destroying vegetative cells of microorganisms of public health significance, and in substantially reducing numbers of other undesirable microorganisms, but without adversely affecting the product or its safety for the consumer.

Scheduled Depuration Process (SDP) - A process which places shellfish harvested from restricted or approved waters into a controlled aquatic environment selected by the processor and approved by the state shellfish control agency as adequate to effectively reduce the level of bacteria and viruses in live shellfish.

Scheduled Heat Shock Process - The process selected by the processor and approved by the state shellfish control agency to heat shock a shellfish species in order to facilitate shucking without adversely affecting the microbial quality or altering the organoleptic characteristics of the species.

Shall - The term used to state mandatory requirements.

Shellfish - All edible species of oysters, clams, mussels and scallops;* either shucked or in the shell, fresh or frozen, whole or in part. (See Part I for a partial listing of common and scientific names.)

Shellstock - Shellfish in the shell.

Shellstock Shipper (SS) - A person who grows, harvests, buys, or repacks and sells shellstock. They are not authorized to shuck shellfish nor to repack shucked shellfish. A shellstock shipper may also ship shucked shellfish.

Should - The term used to state recommended or advisory procedures or identify recommended equipment.

Shucked Shellfish - Shellfish, whole or in part, from which one or both shells have been removed.

---

* Scallops are to be excluded when the final product is the shucked adductor muscle only. [The 1991 ISSC Scallop Committee Report provided a two year period for incorporating whole and roe-on scallops into the NSSP.]

Shucker-Packer (SP) - A person who shucks and packs shellfish. A shucker-packer may act as a shellstock shipper or reshipper or may repack shellfish originating from other certified dealers.

State Shellfish Control Agency (SSCA) - The state agency or agencies having the legal authority to classify shellfish growing waters, to issue certificates for the interstate shipment of shellfish and regulate harvesting, processing and shipping in accordance with the NSSP Manual of Operations, Parts I and II. Foreign shellfish control authorities having effective MOU's or agreements with FDA are considered state shellfish control agencies for the purpose of this manual.

Transaction Record - A form(s) used to document each purchase or sale of shellfish at the wholesale level.

Wet Storage - The temporary storage of shellfish from approved sources, intended for marketing, in containers or floats in natural bodies of water or in tanks containing natural or synthetic seawater.

Satisfactory Compliance

Items within each section of the NSSP Manual Part II are highlighted through the use of **bold print**. These items are identified under the heading **Satisfactory Compliance**. The items shown as satisfactory compliance are the only items in the Manual by which a State Shellfish Program shall be determined to be in compliance with the NSSP guidelines.

(This page is blank.)

Section A

General Administrative Procedures

1. State Laws and Regulations

State laws or regulations shall provide an adequate legal basis for sanitary control of all phases of the harvesting, processing, distribution and shipping of shellfish to interstate markets. This legal authority shall enable one or more departments or agencies of the state to regulate and supervise the source, shipment, labeling and storage of shellfish; the operation of depuration plants; and the shucking, packing, labeling and repacking of shellfish. The state shall also be permitted to certify and decertify interstate shellfish shippers; to conduct laboratory examinations of shellfish; to prevent the sale of unsafe or uncertified shellfish by such legal means as detention, monetary fines, seizure, embargo and destruction; and to suspend interstate shippers certificates in public health emergencies.

Satisfactory Compliance
This item will be satisfied when the state has legal authority to:

a. Certify, inspect, and determine the sanitary rating of the operations of each interstate shipper or processor of shellfish to determine the level of conformity with applicable provisions of Part II of this Manual. Inspection includes the authority to review and copy necessary records to determine whether compliance with the applicable requirements is being maintained.

b. Regulate the shipping conditions and labeling requirements for shellstock to protect against contamination and to provide for accurate source identity. These controls apply to every person who handles shellfish from the point of harvest through each certified shipper and up to the retail point of sale whether or not they are required to have an interstate shellfish shippers certificate.

c. Regulate the buying, selling, processing, packaging, storage, and repacking of shellfish to protect against contamination and product quality degradation, to maintain source and lot identity and integrity, and to provide for proper labeling and packaging.

d. Regulate the depuration of shellstock to prevent illegal diversions, ensure cleansing, protect against recontamination, verify product quality and effectiveness of the SDP, maintain production and product quality records, and provide for proper labeling and packaging.

e. Suspend operations or decertify interstate shellfish dealers on the basis of unacceptable operating and sanitation conditions. An inspection rating score is established for each certified shellfish dealer.

f. Laboratories performing shellfish analyses are evaluated and meet the requirements of Part I, Section B of this Manual.

g. Collect samples and make appropriate bacteriological, chemical, and physical examinations necessary to determine product quality, and monitor the effectiveness and performance of process operations.

h. Prohibit the sale, shipment, or possession of shellfish from unidentified sources; uncertified dealers, unlicensed or unpermitted aquaculturists, or unapproved growing areas; from sources which did not harvest, transport, process, or pack the shellfish in accordance with the requirements of the NSSP and the Federal Food, Drug and Cosmetic Act; or which otherwise caused the shellfish to be unfit for human consumption. Such shellfish shall be detained, condemned, seized, or embargoed. The authority to take enforcement action need not be specific for shellfish and may be included in other state laws.

Public Health Explanation

The NSSP was developed by the 1925 Conference on Shellfish Pollution to meet the specific public health needs resulting from the 1924-1925 typhoid epidemic (1). A principal objective of the NSSP has been to provide a mechanism for health officials and consumers to receive information as to whether lots of shellfish shipped in interstate commerce meet acceptable and agreed upon sanitation and quality criteria. Although these requirements pertain only to interstate shipments, it is recommended that the same requirements be imposed on intrastate operations.

To accomplish this, the NSSP has established criteria and procedures to allow a producing or processing state to "certify" to receiving states that the product from specific dealers has been grown, harvested, transported, processed, and/or shipped in compliance with NSSP criteria. Certification is dependent on a dealer maintaining acceptable operational and sanitary conditions. A rating score is to be established for each certified shellfish dealer based upon an inspection by the state shellfish control agency applying the sanitation criteria of the NSSP and uniform NSSP inspection forms.

The state must have adequate legal authority to regulate the sanitary requirements for harvesting, transporting, shucking-packing, and repacking of shellfish to be shipped interstate. This authority may be either a specific law or a regulation. The success with which the state is able to regulate all components of the shellfish industry provides a measure of the adequacy of the statutory authority.

The unique nature of shellfish as a food also makes it necessary that the SSCA have authority to take immediate emergency action to halt sale and distribution of shellfish without recourse to lengthy administrative procedures. As an example, a state may find it necessary to detain lots of shellfish following reports of illness traced to a certain source of shellfish before confirmatory laboratory analysis can be conducted to document the causative agent. In taking such action, the responsible regulatory agency should be cognizant of the need to use rapid analytical methods for determining status of these highly perishable products.

Periodic revisions of state shellfish laws or regulations may be necessary to cope with new public health hazards and to reflect new knowledge. Examples of changes or developments which have called for revision of state laws include the construction of depuration plants, changes in conservation laws, or the exploitation of a new resource.

2. Certification Procedures To Be Used By States

States shall certify dealers for interstate shipment in accordance with the sanitation and administrative criteria in this Manual. The state shall keep records which will document that the applicable criteria have been applied to all certified dealers. The state shall conduct inspections with such frequency to ensure that satisfactory operational and sanitation conditions are maintained. Dealers having major non-conformities or not meeting and maintaining the minimum sanitation criteria shall not be eligible for inclusion in the ICSSL. NSSP standards also shall be applied to all persons handling shellstock prior to its delivery to the certified interstate shipper.

Satisfactory Compliance
This item will be satisfied when:

a. The state shellfish sanitation control program officials have applied to the FDA for evaluation and have been found to be in conformity with the NSSP before initiating a state shellfish sanitation program or a new program element within an existing state program. The FDA will act on any application submitted by a SSCA within 30 days. If the FDA has not acted within 30 days, the SSCA may proceed with the intended shellfish sanitation program.

b. NSSP requirements are applied to all commercial shellfish harvesters (including aquaculturists); all persons handling the shellfish prior to its delivery to the interstate shipper; all persons engaged in depuration, wet storage, shucking, packing and repacking; and all persons shipping certified shellfish in interstate commerce.

c. All certified dealers, except depuration and wet storage dealers, shall undergo a comprehensive on-site inspection prior to issuance of the initial and thereafter annual certification. The certification period shall not exceed 12 months. This comprehensive on-site inspection shall be conducted by the standardized shellfish plant inspector within ninety (90) days of the application for certification or renewal of certification, show the date of the on-site inspection, the inspector's full name (printed in addition to the signature) and the date of expiration of the inspector's standardization. (This requirement will become effective with the issuance of the January 1, 1994, *Interstate Certified Shellfish Shippers List*.)

d. Only one (1) certification number shall be issued to a dealer per location. Existing facilities having dual certification numbers SHALL be phased out by the SSCA no later than January 1, 1992.

e. Certification is granted only to shippers who meet the following inspection requirements: 1) no *CRITICAL* deficiencies; 2) not more than two (2) *KEY* item deficiencies; and 3) not more than three (3) *OTHER* item deficiencies. After a dealer is certified, unannounced inspections using the NSSP plant inspection form are conducted during periods of operation and at such frequency as necessary to assure that adequate operational and sanitary conditions are maintained. A copy of the completed inspection form and a list of observations for items of non-compliance are provided to the most responsible individual at the firm. The minimum frequency for reinspection is:

i. within 30 days of beginning operation for any dealer certified on the basis of a pre-operational inspection;

ii. at least monthly for a depuration plant;

iii. at least quarterly for shucker-packer and repacker; and

iv. at least semi-annually for other certified dealers.

f. Enforcement actions are taken as follows:

i. When a routine inspection detects a *CRITICAL* deficiency, the deficiency shall be corrected during that inspection or the plant must cease production affected by the deficiency. If the item is not corrected within the specified time, the SSCA will immediately begin actions to withdraw dealer certification. Further, product affected by the *CRITICAL* deficiency shall be controlled to prevent contaminated or adulterated product from reaching consumers. Controls other than product destruction may be considered and approved by the SSCA on a case-by-case basis.

ii. When a routine inspection detects four (4) or more *KEY* item deficiencies, a follow-up inspection is conducted as soon as possible but within 30 days. The follow-up inspection shall determine if the deficiencies have been corrected or are being corrected per the scheduled correction dates noted on the previous inspection report.

iii. When the follow-up inspection of the *KEY* item deficiencies indicates a failure to comply with the correction schedule, the SSCA will immediately begin actions to suspend operations and withdraw dealer certification.

iv. When a routine inspection detects *OTHER* item deficiencies or three (3) or less *KEY* item deficiencies, the deficiencies shall be corrected prior to the next routine inspection.

v. All specific deficiencies, as noted in the narrative section of the inspection report, which are repeated consecutively and are not corrected as scheduled shall be corrected prior to the annual certification. Dealers which fail to correct such deficiencies will not be certified.

vi. When inspections are made of certified shellfish shippers where the SSCA finds non-conformities that present an imminent threat to public health actions are initiated immediately by the SSCA to suspend operations and withdraw certification until a reinspection confirms that appropriate corrections have been made. The control agency also should detain or seize any undistributed lots of shellfish that may have been adulterated, initiate a recall of shellfish distributed intrastate, and notify FDA and receiving state enforcement agencies of product distributions.

vii. When inspections are made of certified shellfish shippers where the SSCA finds major public health deficiencies, action is initiated by the SSCA to suspend or withdraw certification until a reinspection confirms that appropriate corrections have been made.

viii. When a certificate is removed for cause, the SSCA immediately notifies FDA and shellfish control personnel in known receiving states. Member foreign countries shall notify FDA who shall in turn notify receiving states.

g. A certified shellfish dealer whose certificate has been removed for cause may not ship interstate until the SSCA is satisfied that necessary corrections have been made. A recertification is not issued until an inspection establishes that the firm is in substantial compliance with all applicable criteria of this Manual. Upon recertification, the SSCA will notify FDA and known receiving states immediately.

h. Adequate records documenting the degree of compliance with the certification requirements are maintained in a central file for at least three years and made available to the FDA regional office upon request. These records will include:

i. inspection reports of certified dealers;

ii. notification letters and actions taken regarding low sanitation ratings and certification withdrawals;

iii. records of shellfish sample results and follow-up actions taken;

iv. records of complaints or inquiries and follow-up actions taken; and

v. records of administrative hearings.

i. The certifying officer responsible for completing Form FDA 3038b, SHELLFISH CERTIFICATION, forwards the completed form to FDA (HFF-513), 200 'C' Street, S.W., Washington, D.C. 20204 for publication in the monthly listing, and forwards a copy to the appropriate FDA Regional or District Office. The interstate shellfish certificates issued by the state to FDA for publishing should provide the following information:

i. the usual business name and alternative names that should appear on the ICSSL (hereafter called the "List");

ii. a business address where inspections are conducted;

iii. a unique certificate number for each business unit consisting of a one to five digit arabic number preceded by the two letter state abbreviation and followed by the two letter abbreviations for the type of operation the dealer is qualified to perform; shucker packer (SP), repacker (RP), shellstock shipper (SS), reshipper (RS), or depuration processor [depuration] (DP); and

iv. an expiration date that is the same for all firms and is the last day of a month.

j. The state applies the following guidelines in managing the shellfish certificates that they issue.

i. A change in an existing, unexpired certificate, or a withdrawn certificate is made by issuing a corrected certificate.

ii. Shippers are informed by the state certifying officer of the probable date their names will appear on the "List" and should be advised against making interstate shipment prior to that date. If shipments must be made before the appearance of the shipper's name on the "List", the SSCA in the shipper's state will notify the appropriate agency in each of the receiving states and the FDA regional and headquarter's office.

iii. If a state cancels an interstate shellfish shipper certificate, the FDA regional office is notified immediately and a completed Form FDA 3038c, CERTIFICATE CANCELLATION, is mailed to FDA (HFF-513), 200 'C' Street, S.W., Washington, D.C. 20204.

iv. Renewal certificates should be sent to FDA so they are received by FDA's Shellfish Sanitation Branch prior to the date of printing (usually the 15th) of the "List" for the month that the original certificates expire. Certificates will be withdrawn automatically from the "List" on the date of expiration unless new certificates have been received by FDA.

v. Persons desiring to receive copies of the "List" or desiring information concerning the "List" should contact FDA (HFF-513), Shellfish Sanitation Branch, 200 'C' Street, S.W., Washington, D.C. 20204. Recipients will be contacted periodically to determine if they still have use for the "List".

k. State shellfish plant inspectors are provided with appropriate equipment and supplies to conduct adequate inspections of certified shippers.

l. Memoranda of Understanding between state control agencies where responsibilities are divided have been developed which define the responsibilities of each state agency in maintaining adequate sanitary control of the shellfish industry.

Public Health Explanation

State officials who certify dealers must fully comply with the requirements for certification for the process to remain viable. Certification is intended to provide an unbroken chain of sanitation control to a lot of shellfish from the moment of harvest to its sale at the wholesale or retail level. For the certification process to be effective, certified dealers must fully comply with the applicable sanitation requirements pertaining to the type of operation involved.

The minimum plant sanitation and management guidelines for interstate shellfish shippers are described in Sections B-I of this Manual. Only those shellfish firms that meet the guidelines are eligible for certification as Interstate Shellfish Shippers and may be listed in FDA's monthly publication of the ICSSL. Plants having major non-conformities should not be certified and certified plants found to have major non-conformities should have their license or permits suspended or certification canceled. This "List" is mailed to over 6,000 persons to inform them of approved sources of shellfish. Food control officials throughout the United States use the "List" to determine that shellfish offered for sale or used in food service establishments have been produced under the sanitary guidelines of the NSSP. These officials are asked to rely upon the certification process by not holding up shipments or sales of shellfish lots pending examination.

Inspections of certified shellfish dealers should be conducted at such frequency as is necessary to assure compliance with NSSP requirements. The recommended frequency of inspection for shucker packers, repackers, and depuration plants when operating is at least monthly and for shellstock shippers and reshippers at least quarterly. To conduct effective inspections, it is necessary that inspectors have adequate equipment and supplies to measure compliance with applicable requirements. Since the type of equipment and supplies required for an inspection will vary with the type of establishment, it is recommended that a checklist of equipment be developed for each dealer classification.

(This page is blank.)

Section B

Harvesting, Handling and Shipping Shellfish

1. Boats and Trucks

All boats used for harvesting or transporting shellfish, including "buy" boats, and all trucks used for hauling bulk, bagged, containerized, or otherwise packaged shellstock shall be constructed, operated, and maintained so as to prevent contamination, deterioration, or decomposition of the shellfish, and shall be kept clean. Adequately refrigerated trucks shall be used to transport shellstock when the ambient air temperature is such that unacceptable levels of bacterial growth may occur in the shellstock species to be transported.

Satisfactory Compliance
This item will be satisfied when:

a. Decks and storage bins are constructed and located so as to prevent bilge water or polluted overboard water from coming into contact with the shellfish.

b. Bilge pump discharges are located so that pumpage will not contaminate shellfish.

c. Sacks or other containers used for storing shellfish are clean and fabricated from safe materials.

d. Boat decks, truck floors, and storage bins are kept clean with potable water or water from an approved growing area and are provided with effective drainage.

e. Where necessary, effective coverings are provided on harvest boats to protect shellstock from exposure to hot sun, birds and other adverse conditions.

f. Portions of boats or trucks (decks, storage bins, floorbeds, etc.) and all other equipment (shovels, wheelbarrows, rakes, etc.) coming in contact with shellstock during handling or transport from polluted areas to approved areas for relaying are thoroughly cleaned before they are used to transport or handle shellfish from approved areas.

g. Trucks used to transport shellstock are constructed so as to protect the shellfish from contamination and are kept clean.

h. Shellfish are transported in adequately refrigerated trucks when ambient air temperature and time of travel are such that unacceptable bacterial growth or deterioration may occur. Prechilling shellfish before loading onto trucks is recommended and prechilling trucks is required when ambient air temperatures are such that unacceptable bacterial growth or deterioration may occur.

     i. Dogs, cats, or other animals are not permitted on vessels or in vehicles on which shellstock is held.

     j. All owners of "buy" boats and "buy" trucks shall be certified by the SSCA as shellstock shippers.

Public Health Explanation

Precautions exercised in gathering shellfish from approved growing areas may be nullified if shellfish are contaminated with bilge water or polluted overboard water, or in the case of trucks, with contaminated water on the floor or hazardous materials on or adjacent to the shellstock. Also, several investigations have been conducted by States and the FDA regarding shipments of shellfish where product deterioration resulted when shellstock was held or shipped under adverse conditions such as direct sunlight and warm temperatures (13, 14, 15, 16). These studies reaffirm the critical role that adequate shellstock protection and refrigeration plays when ambient temperatures are high. Product deterioration and bacterial growth occurs when shellstock is left exposed for several hours on harvest boats. If this shellstock is transported in trucks without adequate prechilling and in-transit refrigeration, product deterioration continues.

2. Shipping and Receiving

Proper growing area classification and strict adherence to Good Manufacturing Practices (GMPs) are the principal considerations for assessing the safety of shellfish. The SSCA shall not use current bacteriological criteria as the primary basis for embargoing or destroying shellfish shipments.

In cases where an interstate shipment of shellfish is to be monitored, the SSCA shall monitor the acceptability of shellfish shipments within 24 hours of a shipment entering the state in order to determine the source of any problems which may be identified.

     ### Satisfactory Compliance
     This item will be satisfied when the following requirements for interstate shipment of shellfish products are met:

     a. Shellfish shipping conditions shall be considered satisfactory if:

          i. The entire cargo consists only of compatible molluscan shellfish products (oysters and hard clams or soft clams and mussels). Mixed cargo which may include other seafoods in boxes to the same or subsequent destination is acceptable only if the compatible molluscan shellfish products (oysters and hard clams or soft clams and mussels) are protected from contamination by the other cargo through partitioning, horizontal separation or other isolation methods. No other cargo may be placed on or above the shellfish unless all cargo is packed in sealed, crush resistant, waterproof containers.

ii. Shellstock shipments are palletized (except in bulk). Pallets shall be utilized when the conveyances do not have a channeled floor. Mixed cargoes of seafood, other than shellfish, shall be palletized.

iii. The conditions of shipment fully comply with one of the following:

(a) Interstate shipments are made in mechanically refrigerated conveyances maintained at or below 7.2°C (45°F). A suitable time-temperature recording device shall accompany each shipment. The initial shipper shall note the date and time on the recording device. The receiving dealer shall note date and time on the recording device when the shipment is received and doors opened. The recording chart or other record of the time-temperature shall be maintained in the records of the final receiver and presented on request to the state shellfish regulatory agency. An inoperative recording device will be considered as no recording device.

(b) When shipments are limited in duration to four (4) hours or less, shellstock and shucked shellfish products are shipped well-iced in cartons or with other means of acceptable refrigeration, in which case, no thermal recorder is needed. No ice would be an unsatisfactory shipping condition.

iv. The shipper provides security on the load. A tamperproof seal having a serialized identification (ID) number is recommended in which case the ID number shall be recorded on the invoice or bill of lading.

v. Each shipment is accompanied by a shipping document. The receiving dealer shall maintain in his files a copy of the completed document from the shipper which shall be presented on request to the state shellfish regulatory agency. The document shall contain the name, address and certification number of the shipper, the name and address of the major consignee and the kind and quantity of the shellfish product. If the shipment is subdivided to different dealers, records shall be retained so that the original shipment may be traced.

vi. The conveyance in which the shellfish products are shipped is maintained in a sanitary condition while the shellfish are on board.

vii. Containers used for shellstock shipments are clean.

b. Shellfish shall be considered acceptable if:

i. Shipments are properly identified (by tag, bill of lading or label) and cooled, shellstock* to 10°C (50°F) or less, and shucked** product to 7.2°C (45°F) or less, and that there is compliance with all other NSSP conditions of shipment.

ii. Shellstock are alive.

c. Shellfish shall be rejected if:

i. Shipments are without proper identification; or

ii. Shellstock shipments exceed 15.6°C (60°F); or

iii. Shucked product shipments exceed 10°C (50°F); or

iv. Other conditions exist, e.g. decomposition and/or adulteration with poisonous and deleterious substances.

Any of the above constitute prima facie conditions for the SSCA to reject the shipment without testing or further investigation.

d. Shellfish shall be examined bacteriologically if:

i. Shellstock* shipping temperature exceeds 10°C (50°F), but is not higher than 15.6°C (60°F), and shucked** product shipping temperature exceeds 7.2°C (45°F), but is not higher than 10°C (50°F), even though there is proper product identification and compliance with other shipping conditions. The protocol and criteria in Figure 1 (decision tree) are recommended for sampling and adjudicating shellfish shipments.

ii. Deemed necessary by the SSCA at any other time.

---

* Shellstock temperature will reflect internal body temperature.

** Temperature of shucked products will be determined as described in Section F, of this Manual.

Figure 1
Decision Tree

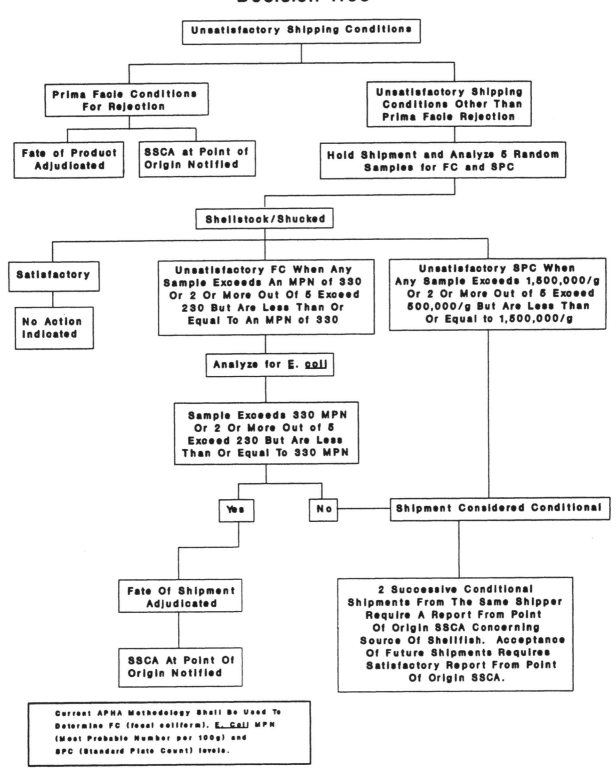

Public Health Explanation

Studies conducted during the period from pre-1925 to 1989 showed that the bacteriological examination of shellfish is an important *tool* in detecting: product mishandling; temperature abuse; and gross errors in growing area classification (1, 3, 4, 5, 6, 11, 13, 14, 15, 16, 17, 18, 19, 20, 21, 22, 23, 24). The studies also demonstrated that shellfish will generally reflect the bacteriological quality of the water in which they have grown. However, this relationship is not consistent. Variation reflects differences in species and product forms and seasonal conditions at the time of harvest (1, 20). Some studies concluded that there is no single uniform bacteriological standard which could be applied to all species of shellfish (20).

Efforts to develop satisfactory bacteriological criteria for interstate shipments of shellfish (especially oysters) as received at the wholesale market level were begun in 1950 (23). During the period from 1950 to 1964, there were many studies conducted to determine the bacteriological changes associated with shellfish harvesting, shucking - packing and marketing. Throughout this period various coliform and plate count standards were developed under the NSSP. However, it wasn't until 1965, that the fecal coliform and standard plate count criteria were applied to all species of shucked oysters at the "wholesale market level" (wholesale market level not defined) (20). In 1968, the NSSP Workshop adopted these criteria, presumably for all species and product forms of oysters, clams and mussels (7).

The majority of studies on microbiological quality of shellfish point up the need to refrigerate shellstock quickly after harvesting and maintain the product below 10°C (50°F) throughout processing, distribution and storage. It should be noted that a study by Cook and Ruple reported in 1989, showed that 10°C (50°F) storage of summer harvested Eastern oyster shellstock from the U.S. Gulf Coast, prevented the multiplication of fecal coliforms and vibrios, including *Vibrio vulnificus* (24). Universally, food control officials consider shellfish as a potentially hazardous food that is capable of supporting rapid and progressive growth of infectious or toxigenic microorganisms. Other foods in this category are milk, milk products, eggs, meat, poultry and fish. Generally, FDA recommends that potentially hazardous food be held at 7.2°C (45°F) or below, and if large volumes are involved in processing, methods be employed to rapidly cool the product to an internal temperature of 7.2°C (45°F) within four hours (20).

3. Washing of Shellstock

Shellstock should be washed reasonably free of bottom sediments and detritus as soon after harvesting as practicable. The primary responsibility for washing rests with the harvester. Water used for shellstock washing shall be obtained from an approved growing area, or from other sources approved by the SSCA.

Satisfactory Compliance
This item will be satisfied when:

a. Shellstock are washed reasonably free of bottom sediments and detritus as soon after harvesting as is feasible. Washing of naturally clean shellstock is not necessary. Shellstock are washed at the time of harvest except when this is not feasible because of harvesting methods or climatic considerations, and it is feasible for the shipper or processor to assume this responsibility.

b. Water used for washing shellstock is obtained from an approved growing area, or from other safe sources approved by the SSCA.

c. If shellstock is washed with systems which recirculate the wash water, the following requirements shall be met:

   i. All plans for construction or remodeling of a recirculating shellstock washing system shall be reviewed and approved by the SSCA prior to commencing construction.

   ii. A water treatment and disinfection system shall be utilized in the recirculated system, which provides an adequate quantity and quality of water for the shellstock washing operation. The treatment shall not leave residues that are not Generally Recognized As Safe (GRAS). Treated water utilized immediately after disinfection for shellstock washing shall have no detectable levels of coliform organisms as measured by the standard five tube MPN test for drinking water (25). Treated water shall be tested daily by a laboratory approved by the SSCA.

   iii. For water receiving UV disinfection, turbidity shall not exceed 20 nephelometric turbidity units (NTU's) measured in accordance with *Standard Methods for the Examination of Water and Wastewater* (25).

   iv. Disinfection units shall be cleaned, serviced, and tested as frequently as is necessary to assure effective disinfection.

   v. Storage tanks and related plumbing shall be fabricated of safe material and shall be easily cleanable. Tanks shall be constructed so as to be easily accessible for cleaning and inspection, to be self-draining, and to meet food-contact surface requirements. Plumbing shall be designed and installed so that cleaning and sanitizing will be effective. Tanks shall be cleaned and sanitized on a regular schedule to prevent contamination of the tank and water.

d. Shellstock shall not be placed in containers of stagnant water for the purpose of washing or loosening sediment.

Public Health Explanation

Several studies have established that some pathogenic *Vibrio* species and other autochthonous bacteria may be present in marine sediments throughout the year (26, 27, 28). One study of *Vibrio* species and *Aeromonas hydrophila* in sediments of Apalachicola Bay, Florida, routinely detected *V. parahaemolyticus*, *V. alginolyticus*, and *A. hydrophila* and during some portions of the year at relatively high levels (up to 46,000 organisms per gram). Additionally, *V. vulnificus*, *V. cholerae*, *V. fluvialis* were detected at levels up to 2,400 organisms per gram of sediment (27).

Vibrio species can also survive on inadequately cleaned equipment in a processing plant (29). Washing sediments from shellstock at the time of harvest helps to protect the shellfish and the processing equipment from becoming contaminated. Washing shellstock also helps to prevent quantities of mud and other bacteria from being mixed with the shucked shellfish, thereby contributing to high bacterial counts in the finished product (17, 30). Muddy shellstock also makes it difficult to maintain shucking rooms in a clean, sanitary condition.

Water used for shellstock washing should be of good sanitary quality, to avoid possible contamination of the shellstock. There are instances when shellstock washing by the harvester might introduce a sanitary hazard because of the possible tendency of the harvester to wash the shellstock with polluted water from a harbor area, rather than with clean water from a growing area. Therefore, SSCA may waive the requirement for shellstock washing by the harvester when there are climatic, technical, or sanitary reasons for such action. In such event, the processor becomes responsible for washing shellstock.

4. Disposal of Body Wastes

Body wastes shall not be discharged overboard from a boat used in the harvesting of shellfish, or from "buy" boats while in areas from which shellfish are being harvested. The appropriate state agency, when necessary, shall specify the device and practices necessary to eliminate the overboard discharge of body wastes from boats used in harvesting of shellfish. Educational materials should be provided to all boat owners concerning the public health significance and dangers in discharging body wastes overboard.

Satisfactory Compliance
This item will be satisfied when:

a. Body wastes are not discharged from harvest or buy boats while in an area approved for shellfish harvesting.

b. Portable toilets, if provided, are used only for the purpose intended, and are so secured and located as to prevent contamination of the shellfish by spillage or leakage.

c. The contents of portable toilets are emptied only into an approved sewage disposal system, and portable toilets are cleaned before being returned to the boat. (Facilities used for cleaning food-processing equipment may not be used for this purpose.)

d. The SSCA has an education program which provides all boat owners with information concerning the public health dangers in discharging body wastes overboard in approved shellfish growing areas.

Public Health Explanation
It is necessary to protect the shellfish from pollution by disease-causing organisms that may be present in body wastes discharged from boats. This item is intended to protect the shellfish from chance pollution during harvesting. The likelihood of body wastes being discharged from boats will be considered in evaluating the sanitary quality of the harvesting area. If discharges are not adequately controlled, the area cannot meet the classification requirements for an approved harvesting area.

5. Shellstock Packaging and Identification
Shellstock containers shall be clean and fabricated from safe materials. Each licensed harvester or certified dealer shall securely affix to each container of shellstock a tag or label approved by the appropriate SSCA and bearing all information necessary to trace the shellfish back to a specific area and a particular harvester or group of harvesters. When shipments are transported in bulk, the harvester shall provide a transaction record containing the required information. The SSCA may designate either the harvester or the certified dealer as legally responsible for shellstock identification.

Satisfactory Compliance
This item will be satisfied when:

a. Each shellfish harvester has a valid license as required in Part I, Section F.1.

b. Sacks, boxes, and other shellstock packing containers are clean and fabricated from safe materials.

c. Both the harvester or aquaculturist and initial certified dealer are legally required to identify shellstock in accordance with the provisions of this Manual.

d. Each harvester or aquaculturist and each certified dealer affixes an approved, durable, waterproof tag of minimal size — 6.7 by 13.3 cm (2 5/8 by 5 1/4 inches) — to each container of shellstock; for harvesters this shall be done prior to landing unless the harvest has occurred at more than one harvest location or aquaculture site, then each container shall be tagged at the harvest location or aquaculture site; for dealers this shall be done prior to shipment. In the case where a certified dealer is also the harvester, that dealer's tag may also be used as the harvester's tag.

e. The harvester's tags shall contain legible information arranged in the specific order as follows:

  i. a place may be provided where the dealer's name, address and the certification number assigned by the SSCA may be added;

  ii. the harvester's identification number as assigned by the SSCA;

  iii. the date of harvesting;

  iv. the most precise identification of the harvest location or aquaculture site as is practicable (e.g. Long Bay, Smith's Bay, or a lease number);

  v. type and quantity of shellfish; and

  vi. the following statement will appear in bold capitalized type "THIS TAG IS REQUIRED TO BE ATTACHED UNTIL CONTAINER IS EMPTY OR RETAGGED AND THEREAFTER KEPT ON FILE FOR 90 DAYS."

f. The dealer's tags shall contain legible information arranged in the specific order as follows:

  i. the dealer's name, address, and the certification number assigned by the SSCA;

  ii. the original shipper's certification number including the state abbreviation;

  iii. the date of harvesting;

  iv. the most precise identification of the harvest location or aquaculture site as is practicable; this identification *must* include at least the state (initials) in which the shellfish were harvested and the designated harvest area with that state as assigned by the SSCA of the producer state. If harvest areas have not been indexed by the SSCA, then an appropriate geographical or administrative designation must be used (e.g. Long Bay, Decadent County, lease number, bed or lot number);

  v. type and quantity of shellfish; and

  vi. the following statement will appear in bold capitalized type "THIS TAG IS REQUIRED TO BE ATTACHED UNTIL CONTAINER IS EMPTY AND THEREAFTER KEPT ON FILE FOR 90 DAYS."

g. When shellstock are sold in bulk, the harvester and/or certified dealer shall provide a transaction record prior to shipment. This record from harvesters shall contain information identical to that required in Section B.5.e with the addition of the name of the consignee in the case of bulk shipments. This record from dealers shall contain information identical to that required in Section B.5.f with the addition of the name of the consignee in the case of bulk shipments.

Public Health Explanation

Licensing of each person who harvests shellfish for sale to a certified dealer is an important control measure to help protect against illegally harvested shellfish and to help maintain accurate source identity records. Harvesters must provide information necessary to create a record of the origin, quantity, and date of harvest which can be used to trace lots of questionable shellstock back to the source(s). Investigation of disease outbreaks can be severely hindered if the source of the shellfish cannot be readily identified. This can result in shellfish from the unacceptable source continuing to be used and continuing to cause illness. Health authorities may be forced to close safe areas, to ban safe shipments or to seize safe lots as a public health precaution if the source of contaminated shellfish cannot be accurately and rapidly determined.

(This page is blank.)

Section C

Wet Storage

Temporary wet storage of live shellstock in nearshore floats, baskets, or sacks, and onshore in tanks has been practiced worldwide for many years. There are three primary reasons for wet storage. First, shellfish can be harvested from remote areas at convenient times and held live for brief periods at locations close to the point of sale in containers from which they can be easily retrieved. Wet storage also may be used to desand those shellfish species which tend to accumulate sand in their mantles and gills thus making them more palatable. Thirdly, wet storage may be used to increase palatability by increasing salt content of shellfish harvested from low salinity waters. These requirements do not apply to transplant operations where shellfish are moved to new growing areas for conditioning or resource management.

1. Source of Shellfish

Shellfish for wet storage shall be harvested only from approved or conditionally approved areas and shall be harvested, identified and shipped in accordance with applicable requirements of this Manual.

> ### Satisfactory Compliance
> **This item will be satisfied when the shellfish are harvested from approved or conditionally approved areas and are harvested, shipped and identified according to the requirements of Section B.**

Public Health Explanation

The purposes of wet storage are the temporary storage of approved shellfish, desanding and improving palatability. Wet storage facilities are not designed and operated to increase safety of the shellfish. Therefore, all controls pertaining to shellfish for direct consumption should be applied.

2. Storage Facilities

Effective control measures shall be established and implemented to protect shellfish in wet storage against contamination. Necessary control measures include reviewing plans giving the location of storage areas or floats, reviewing design and operating procedures for onshore facilities, periodically inspecting wet storage facilities, and evaluating water supplies for compliance with requirements of this Manual.

> ### Satisfactory Compliance
> **This item will be satisfied when:**
>
> **a. Each wet storage site or facility is evaluated and approved annually by the SSCA on the basis of an evaluation of the nearshore site or the facilities plan and operating**

procedures for an onshore operation submitted by the dealer and an inspection of the storage site or facility. Factors to be considered include but are not limited to the following:

i. The sanitary survey of the nearshore storage site with special consideration of potential intermittent sources of pollution;

ii. The location of nearshore storage sites and floats;

iii. A plan giving the design of the onshore storage facility, source of water to be used for wet storage, and details of any water treatment system. All plans for construction or remodeling of onshore wet storage facilities shall be reviewed and approved by the SSCA prior to commencing construction; and

iv. The purpose of the wet storage operation, such as holding, conditioning, or salinization, and any species specific physiological factors that may affect design criteria.

b. Wet storage is practiced only in strict compliance with the provisions in the written approval for the wet storage operation given by the state shellfish control agency.

c. Nearshore areas used for wet storage meet the NSSP approved area bacteriological criteria at all times shellfish are being held for direct marketing.

d. While awaiting placement in a wet storage facility or area, shellfish shall be protected from physical or thermal abuses which may reduce the effectiveness of the wet storage process.

e. Each onshore wet storage facility meets the following design, construction and operating requirements:

i. Effective barriers shall be provided to prevent entry of birds, animals, and vermin into the area.

ii. Floors, walls, and ceilings shall be constructed; and lighting, plumbing, and sewage disposal systems shall be installed to comply with applicable provisions of Section D.

iii. Storage tanks and related plumbing shall be fabricated of safe material and shall be easily cleanable. Tanks shall be constructed so as to be easily accessible for cleaning and inspection, to be self-draining, and to meet food-contact surface requirements. Plumbing shall be designed and installed so that cleaning and sanitizing can be conducted on a regular schedule, as specified in the operating procedures, to prevent contamination of the tank and water.

iv. Storage tank design, dimensions and construction are such that adequate clearance between shellfish and the tank bottom shall be maintained.

v. Shellfish containers, if used, shall have a mesh-type construction which allows water flow to all shellfish in the containers.

vi. Process water does not adversely affect the sanitary quality of the stored shellfish.

vii. In recirculating wet storage systems or systems utilizing source water from less than approved areas, a water treatment system shall be utilized which provides an adequate quantity and quality of water to carry out the intended purpose of the wet storage operation. The treatment shall not leave residues that are not Generally Recognized As Safe (GRAS) or that may interfere with the process. The quality of the source water prior to treatment shall be no less than for a restricted area as defined in Part I, Section C.5. A water sampling schedule is included in the firm's operating procedures and water is tested according to the schedule. If the water source for a flow-through system is from an area classified as less than approved (e.g., restricted), daily samples of the treated water shall be required. Treated water entering the wet storage tanks shall have no detectable levels of total coliform bacteria as measured by any of the approved NSSP laboratory tests (25) listed in Part I, Appendix J, for the analysis of total coliform bacteria in seawater.

If a recirculating water system is utilized, a water treatment effectiveness study shall be conducted to determine that the system will consistently produce water with no detectable levels of total coliform bacteria under normal operating conditions. Thereafter, the frequency of routine sampling for such systems shall be at least weekly. In the event that a single sample contains any detectable levels of total coliforms, as measured by an approved NSSP laboratory test (Part I, Appendix J) for total coliforms, daily sampling shall be instituted immediately until the problem is identified and demonstrated to be corrected.

The wet storage water treatment effectiveness study shall consist of five (5) sets of three (3) samples from each treatment unit. Samples shall be collected at the outlet of the water treatment unit or at the point where the treated water enters the holding tank. In addition, one (1) sample per day of untreated water will be collected and analyzed. These samples shall be collected daily for five (5) days and analyzed by an approved NSSP laboratory test (Part I, Appendix J) for total coliforms. Any positive sample of treated water shall constitute a failure of the effectiveness study, which shall be repeated after corrections are made.

A confirmation for the treatment effectiveness is required whenever more than ten (10) percent of the water volume is added from a less than approved area or each time new UV bulbs are installed. A set of three (3) samples of treated water and one (1) sample of untreated water shall be collected in one (1) day to reaffirm the effectiveness study.

When corrections are required because positive sample results are found in any sample of treated water, the effectiveness of the correction shall be demonstrated through the collection of three (3) samples of treated water and one (1) sample of untreated water in one day following the correction.

viii. For water receiving UV disinfection, turbidity shall not exceed 20 nephelometric turbidity units (NTU's) measured in accordance with *Standard Methods for the Examination of Water and Wastewater* (25).

ix. Water from approved growing areas may be used in wet storage tanks without disinfection if the system is of continuous flow-through design and provided that the nearshore water source used for supplying the system meets the NSSP approved area bacteriological criteria at all times that shellfish are being held for direct marketing.

x. Shellfish shall be thoroughly washed with water from an approved source and culled to remove dead, broken, or cracked shellfish prior to wet storage in tanks. Due to the adverse effects of culling on mussel physiology, culling of mussels may be done after wet storage, subject to approval by the SSCA as specified in the operating procedures.

xi. Shellfish from different harvest areas shall not be commingled during wet storage in tanks. If more than one harvest lot of shellfish are being held in wet storage at the same time, the identity of each harvest lot shall be maintained through the wet storage process.

xii. Bivalve mollusks shall not be commingled with other species in the same tank. Where multiple tank systems use a common water supply system for bivalve mollusks and other species, process water shall be effectively disinfected prior to entering tanks containing the bivalve mollusks.

xiii. Salt added to increase salinity or produce synthetic seawater is free of any levels of poisonous or deleterious substances which may contaminate the shellfish.

f. Complete and accurate records shall be maintained for at least 90 days which will enable a lot of shellstock to be traced back to the wet storage location. Results of water samples and other required tests shall be maintained for a minimum of 2 years.

Public Health Explanation

The types, locations, and purposes of wet storage operation are highly variable and may range from temporary storage nearshore in approved areas to onshore tanks using recirculating, synthetic seawater for the purpose of desanding and salt uptake. Consequently, it is not possible to provide detailed guidelines in this Manual, and it is necessary for each separate operation to be developed and evaluated on its own merit with respect to overall program guidelines.

Removing shellfish from growing beds to storage areas close to shore and habitations may subject such shellfish to constant or intermittent pollution. Shellfish in wet storage tanks are similarly subjected to pollution if the tank water is obtained from a polluted source. An example of such contamination is the 1956 outbreak of infectious hepatitis in Sweden (691 cases) attributed to oysters contaminated in a wet storage area (30).

Shellfish on floats nearshore may be more directly exposed to potential contamination from boats and surface runoff than are shellfish in their natural growing areas. Therefore, particular emphasis should be placed on a sanitary survey of the vicinity to assure that chance contamination does not occur (31).

Careful consideration must be given to designing and operating onshore wet storage tanks to assure that shellfish are not contaminated during holding or do not die from physiological stresses such as low dissolved oxygen and unsuitable temperatures or salinity (32, 33). Excessive mud on the shells and dead shellfish may increase bacterial loads in the tanks and lead to increased microbial levels in the shellfish during storage. Hence, washing and culling the shellfish prior to storage is essential.

Proper hydraulic design of the tank is important to assure an adequate quantity and quality of water with minimum turbulence at suitable temperatures to achieve the intended purpose of the storage operation. Inadequate flow or "dead spots" can lead to oxygen deficiency and shellfish mortality if the shellfish are physiologically active. Minimum turbulence will permit feces and pseudo feces generated by active shellfish to settle out without being resuspended and ingested. Tanks fabricated with safe material which are easily cleanable will prevent possible adulteration with chemicals migrating from the tank into the water and will facilitate cleaning and sanitizing.

Commingling of bivalve mollusks with other species in tanks may subject the bivalve mollusks to contamination from pathogenic organisms from the non-molluscan animals. Fish, crabs, lobsters, and other marine species may be harvested from polluted areas and may have ingested pathogens or accumulated them on their body surfaces. Therefore, holding such animals in the same tank with bivalve mollusks presents a risk of cross contamination. This risk can be avoided by using separate tanks for non-bivalve molluscan species. Where the same water is used for all tanks, effective disinfection must be provided prior to entering the tank holding the bivalve species.

Section D

Shucking and Packing Shellfish

1. Plant Location, Grounds and Arrangements

Plants in which shellfish are shucked and packed or repacked should be located so that they will not be subject to flooding by high tides. If plant floors become flooded, processing shall be discontinued until after waters have receded and the building is adequately cleaned and sanitized. The grounds about a plant in which shellfish are processed shall be free from conditions which may result in contamination of the shellfish at any time during processing and storage. Buildings and structures shall be suitable in size, construction, and design to facilitate proper maintenance and operation. Protection against cross-contamination of shucked products shall be provided by conducting shucking and packing operations in separate locations or in separate rooms. A shucked-stock delivery window shall be used where necessary to provide protection. Packing rooms shall be of sufficient size to permit sanitary handling of the product and thorough cleaning of the equipment. Separate storage facilities shall be provided for storing employee's street clothing, aprons, and personal articles.

Satisfactory Compliance
This item will be satisfied when:

a. Processing and shipping facilities are so located that they will not be subjected to flooding by ordinary high tides. If plant floors are flooded, all operations are discontinued until waters have receded and the building is cleaned and sanitized.

b. The grounds about a plant under the control of the operator are free from conditions which may result in the contamination of shellfish, including, but not limited to, the following:

> **i. Improperly stored equipment, litter, waste, refuse, and uncut weeds or grass within the immediate vicinity of the plant buildings or structures that may constitute an attractant, breeding place, or harborage for rodents, insects, and other pests;**

> **ii. excessively dusty roads, yards, or parking lots that may constitute a source of contamination in areas where shellfish are exposed; and**

> **iii. inadequately drained areas that may contribute contamination to shellfish products through seepage or foot-borne filth and by providing a breeding place for insects or microorganisms.**

c. If the plant grounds are bordered by grounds not under the operator's control of the kinds described in paragraphs b.i. through iii. of this section, care must be

exercised in the plant by inspection, extermination, or other means to effectively exclude pests, dirt, and other filth that may be a source of shellfish contamination.

d. Shucking and packing operations are carried out in separate rooms or in sufficiently separated areas so that there is no likelihood of the shucked product or packing room equipment being contaminated by splash or by other means from adjacent areas.

e. Other manufacturing operations which could result in contamination of shellfish are not conducted in the same area where shellfish are being processed.

f. Sufficient space is provided to place equipment and to store materials in order to assure sanitary operations and safe shellfish production. Fixtures, ducts, and pipes are not suspended over food processing or storage areas or over areas in which containers or utensils are stored or washed. Aisles or working spaces between equipment, and between equipment and walls, are unobstructed and sufficiently wide to permit employees to perform their duties without contaminating shellfish or food-contact surfaces with clothing or personal contact.

g. Where shucking and packing operations occur in separate rooms, a delivery window or area is provided so that shuckers do not have to enter the packing area. The delivery window is equipped with a corrosion resistant shelf constructed of smooth, easily cleanable materials which can be effectively sanitized.

The shelf drains toward the shucking room and, if necessary, is curbed on the packing-room side.

h. Storage facilities are provided outside the food processing area which have adequate capacity for storing clothing, aprons, gloves, and other personal articles of the employees.

Public Health Explanation

The nature of the shucking operation is such that the shucked shellfish meats require protection from undesirable microorganisms, chemicals, filth, or other extraneous materials. This protection is achieved by properly selecting the plant location, maintaining the plant grounds, and using physical barriers or space separation to prevent contamination of shellstock and shucked products.

It is normal during the shucking operation for the shuckers' clothing to become very soiled. If shuckers enter the packing room, shucked stock, cans, and other equipment may become contaminated. A delivery window has proven to be an effective mechanism to help keep shuckers out of the packing room. Rooms or lockers should be provided for clothing, aprons, and gloves to eliminate the tendency to store such articles on the shucking benches or in packing rooms, where they interfere with plant clean-up and operation.

2. Dry Storage and Protection of Shellstock

Shellstock in dry storage shall be protected from contamination and maintained at temperatures necessary to minimize microbial growth. Shellstock from different sources shall be separated as necessary to avoid commingling and aid in maintaining source identity during shucking and repacking operations.

Satisfactory Compliance
This item will be satisfied when:

a. The storage-area floor is impervious to water, is free of cracks and uneven surfaces that create sanitary problems and interfere with drainage, and is graded to assure complete and rapid drainage of water away from the shellfish.

b. Walls and ceilings of shellstock storage rooms and hoppers are smooth, light colored, and constructed of material that will not deteriorate under repeated washing.

c. Shellstock storage areas are constructed so they will not receive floor drainage from other portions of the plant. Shellstock is stored in such a protected location or at an adequate height off the floor to prevent them from coming into contact with water which might accumulate on the floor or from splash by foot traffic. Shellstock storage areas should not serve as an entry way to other areas of the establishment. Shellstock storage areas are protected against sewage backflow by installation of an adequate air gap in the waste line or by providing a separate drain system.

d. Conveyances or devices used in transporting shellstock are constructed, maintained and handled in a way that prevents contamination of the shellstock. Where overhead monorails or conveyors are used, hydraulic fluid or lubricants do not leak or drip onto the shellfish or conveyance surfaces.

e. Shellstock in dry storage shall be protected from contamination and maintained at a temperature of 10°C (50°F) or less. Shellstock shall be chilled to 10°C (50°F) or less within 24 hours of receipt at the facility unless the product is shucked within 24 hours of harvest. At points of transfer, such as the loading dock or in the plant prior to shucking, shellstock is protected from contamination and is not permitted to remain under adverse temperature conditions for prolonged periods.

f. Mechanical refrigeration facilities are adequate for the intended purpose and are equipped with automatic temperature regulating control and indicating thermometers installed to accurately measure the temperature within the storage compartments.

g. Shellstock from different harvest areas within the same state may be commingled if the SSCA permits and develops a written commingling management plan. The objective of any state management plan is to minimize the commingling dates of harvest and harvest areas. The plan shall limit the practice of commingling to primary dealers and to shellstock purchased directly from harvesters from specific areas as defined by the SSCA. The plan shall further define how the commingled shellfish will be identified. The SSCA will define the primary level dealer.

h. Only live shellstock are shucked or sold.

Public Health Explanation
If shellstock are stored where polluted ground or surface water or floor drainage can accumulate, the shellstock may become contaminated. Shellstock may also be contaminated by birds, animals, and vermin (including, but not limited to, insects and rodents).

When shellstock are not refrigerated during prolonged storage, the quality of the product will deteriorate as bacterial levels increase. Studies of shellstock handling have demonstrated that inadequate refrigeration will accelerate the growth of spoilage organisms as well as *Vibrio* species which have been implicated in shellfish-related disease outbreaks (14, 15, 33, 34). Inadequate storage temperatures will also result in decreased shelf life. Other studies have demonstrated that *Vibrio* species are sensitive to cold and rapidly become inactive at low storage temperatures (35). Due to species variations in response to temperature, specific storage criteria should be developed for each species which will minimize bacterial growth and product deterioration.

Separating shellstock from different sources is useful to maintain lot separation during shucking and repacking operations and to assist in tracing shellfish back to the source when questions of quality or safety arise.

3. Floors
Floors shall be constructed of materials impervious to water, be graded to drain quickly, be easily cleanable and be maintained in good condition.

Satisfactory Compliance
This item will be satisfied when:

a. The floors of all rooms in which shellfish are shucked or packed, or in which utensils are washed, are constructed of easily cleanable, corrosion-resistant, impervious material.

b. The floor surface is graded to drain and is free from cracks and uneven surfaces that create sanitary problems and interfere with drainage. Junctions between floors and walls are impervious to water. Floors are maintained in good repair.

Public Health Explanation
Properly graded floors, of durable, impervious material, maintained in good condition, permit rapid disposing of liquid and solid wastes, and facilitate easy cleaning of the plant.

4. Walls and Ceilings

The interior surfaces of rooms in which shellfish are shucked or packed, or in which utensils are washed, shall be smooth, washable, light-colored, and kept in good repair.

Satisfactory Compliance
This item will be satisfied when interior surfaces are constructed of easily cleanable, corrosion resistant, impervious, and light-colored materials. They are constructed and maintained so as to prevent contamination of shellfish during holding or processing.

Public Health Explanation
Smooth, washable walls and ceilings are more easily kept clean and are, therefore, more likely to be kept clean. A light-colored paint or finish aids in the distribution of light and in the detection of unclean surfaces. Clean walls and ceilings are conducive to sanitary handling of shellfish.

5. Insect and Vermin Control Measures

Safe and effective measures shall be used to prevent the entry of insects, rodents, and other vermin and to kill and capture insects and vermin which enter the plant despite other control measures.

Satisfactory Compliance
This item will be satisfied when:

a. Openings to the outside are effectively protected against the entry of vermin and insects by tight-fitting, self-closing doors, and closed windows; or effective screening; or controlled air currents; or other means. Screening material is not less than 15 mesh per 2.54 cm (15 mesh per inch).

b. Insects, rodents, and other vermin are not present.

c. Necessary internal insect and vermin control measures are used, and such measures are in compliance with all state and federal regulations. The use of insecticides and rodenticides is permitted only under such precautions and restrictions as will prevent the contamination of shellfish or packaging materials with illegal residues, and cause no health hazards to employees.

Public Health Explanation
Controlling flies, cockroaches, and other insects may prevent shellfish and food-contact surfaces from becoming contaminated with disease organisms. Controls should be directed at preventing the entrance of insects, rodents, and other vermin into the building, and at depriving them of food, water, and shelter.

6. Lighting

Safe and adequate lighting shall be provided in all handwashing areas; all dressing, locker, and toilet rooms; all areas where shellfish are processed and stored; all areas where equipment and utensils are cleaned; and all areas where containers and other packaging materials are stored.

Satisfactory Compliance
This item will be satisfied when:

a. Lighting is adequate to allow the intended operation to be performed.

b. Light bulbs, fixtures, and skylights or other glass suspended over exposed shellfish are of a safety-type or are otherwise protected to prevent food contamination in case of breakage.

Public Health Explanation

Adequate lighting encourages and facilitates keeping rooms, equipment, and the product clean by making dirt and insanitary conditions conspicuous. Shielded light fixtures help protect the food, equipment and employees from glass fragments should the fixture break.

7. Heating and Ventilation

Working rooms shall be adequately ventilated and heated or cooled when necessary. Operation of cooling, heating or ventilating equipment shall not create conditions that may cause shellfish to become contaminated.

Satisfactory Compliance
This item is satisfied when:

Adequate ventilation is provided to minimize odors, noxious fumes, vapors or condensation (including steam) in areas where food may become contaminated. The ventilation system, heating, or cooling do not create conditions that may cause food to become contaminated by airborne contaminants.

Public Health Explanation

Comfortable working conditions increases the efficiency of the workers, and may promote sanitary practices. Adequate ventilation reduces condensation and aids in retarding the growth of mold. Excessive temperatures also promote growth of spoilage microorganisms in shellfish and on food-contact surfaces.

8. Water Supply

The water supply shall be properly constructed and protected, be easily accessible, adequate, and of a sanitary quality.

Satisfactory Compliance
This item will be satisfied when:

a. Potable water is from a safe source, is protected from contamination, and the water supply system is constructed, maintained, and operated according to law.
b. Running water is provided at a suitable temperature and adequate pressure in all areas where needed to process food; clean equipment, utensils or containers; and supply sanitary facilities.

c. A state regulatory agency collects water samples for bacteriological examination at not less than semi-annual intervals if the supply is from a private source. In addition, samples for bacteriological examination are collected from all new private sources of supply before they are used, and from repaired supply facilities after they have been disinfected. The state regulatory agency has the option of requiring such samples to be collected by approved persons for bacteriological examination at a state approved or certified laboratory. Bacteriological examination shall be made in conformity with the standard methods recommended by the APHA (25).

d. Hot and cold water are provided at each sink compartment, except that warm water is acceptable at handwashing sinks.

e. Steam used in contact with food or food-contact surfaces is free from any materials or additives other than those specified in 21 CFR 173.310 (36).

Public Health Explanation
The water supply should be accessible in order to encourage its use in cleaning operations. It should be adequate to ensure proper washing, rinsing, and sanitizing of the equipment. Finally, it should be safe and sanitary to avoid contamination of food-contact surfaces and the product.

9. Plumbing and Related Facilities
Plumbing shall be adequately designed, installed, and maintained to supply potable water to the plant and to remove sewage and floor drainage from the plant. There shall be sufficient handwashing and toilet facilities conveniently located to promote sanitary employee practices.

Satisfactory Compliance
This item will be satisfied when:

a. Plumbing is installed in compliance with applicable state laws, and is of adequate size and design to:

   i. carry sufficient quantities of water to required locations throughout the plant;

   ii. properly convey sewage and liquid disposable waste from the plant;

iii. ensure that the water supplies and food-contact surfaces are not contaminated as a result of an inadequate plumbing system;

iv. provide adequate floor drainage in all areas where floors are subject to flooding-type cleaning or where normal operations discharge water or other liquid waste on the floor.

b. There are no cross connections between the approved pressure water supply and water from a non-approved source, and there are no fixtures or connections through which the approved pressure supply might be contaminated by backsiphonage. Adequate devices approved by a regulatory agency are installed to protect against backflow and backsiphonage at all fixtures and equipment where the air gap between the water supply inlet and the fixture's flood level rim is less than twice the diameter of the water system inlet. All submerged inlets, including hoses attached to faucets, are equipped with a backflow prevention device. If booster pumps are connected directly to the potable water supply, the pumps are equipped with a low pressure cut-off device or equivalent method to prevent backsiphonage.

c. Handwashing facilities are adequate in number and size for the number of employees, are convenient to the work areas, and are so located that the person responsible for supervision can readily observe that employees wash their hands before beginning work and after each interruption. There is at least one handwashing facility in the packing room for use by packing room workers. Three-compartment sinks used for washing and sanitizing equipment and utensils are not used for handwashing.

d. Handwashing facilities are provided with hot water of at least 37.8°C (100°F) from either a controlled temperature source with a maximum temperature of 46.1°C (115°F), or from a hot and cold mixing or combination faucet. Steam-water mixing valves or steam-water combination faucets are not acceptable.

e. A supply of hand-cleansing soap or detergent, and, where appropriate, sanitizers are available at each handwashing facility. A supply of disposable towels in a suitable dispenser or a suitable hand-drying device that provides heated air are conveniently located near each handwashing facility. Common towels are prohibited. If disposable towels are used, easily cleanable waste receptacles are conveniently located near the handwashing facilities. Handwashing signs, in language understood by the employee, are posted in toilet rooms and near handwashing facilities. Handwashing facilities, hand-drying devices, and all related facilities are kept clean and in good repair.

f. An adequate number of conveniently located, separate toilets are provided for each sex, except that, where satisfactory toilets are located nearby, separate facilities are not required for family shucking or when the plant has fewer than 10 employees. Toilet-room doors are tight-fitting, self-closing, and do not open directly into a processing area. Toilet rooms are kept clean and in good repair. A supply of toilet paper in a suitable holder is available in the toilet rooms. Air vents are screened or have self-closing louvers.

g. No drainpipes or wastepipes are located over food processing or storage areas, or over areas in which containers are stored or washed.

Public Health Explanation

The organisms causing typhoid fever, hepatitis, and other gastrointestinal diseases may be present in the body discharges of cases or carriers, and thus be present in the drain pipes in the plants. Correctly installed plumbing protects the water supplies against cross connections and backsiphonage. A safe water supply in a plant contributes to product purity and to the safety of the workers.

Adequate toilet and handwashing facilities, including running water, soap, and sanitary drying facilities also are essential to the personal cleanliness of the workers. The posting of a handwashing sign is necessary to remind plant employees of this important public-health practice.

10. Sewage Disposal

Sewage shall be discharged into an adequate sewerage system or disposed of through other effective means. Where private sewage disposal systems are utilized, they shall be constructed and maintained according to state and local laws. Privies are acceptable only where water carriage systems are not feasible. The sewerage system shall be constructed and maintained in order that waste will be inaccessible to flies and rodents and shall not provide a source of contamination.

Satisfactory Compliance
This item will be satisfied when:

a. Sewage is discharged into public sewers wherever possible.

b. Private sewage disposal facilities are constructed and operated to ensure that human excreta or sewage are inaccessible to insects and animals, and that contamination by any other means does not occur. Privies approved by the SSCA are acceptable only when water carriage systems are not feasible.

Public Health Explanation

Disease-causing microorganisms may be present in the body discharges of employees that are cases or carriers of communicable disease organisms. When sewage disposal facilities are of a satisfactory type, there is less possibility that the shellfish being processed may become contaminated with fecal material carried by flies, rodents, or by other means. Proper maintenance will ensure that sewage does not leach to the surface from a subsurface system.

Non-water-carriage sewage-disposal systems should be of a sanitary type so that excreta are not accessible to flies or rodents.

11. Poisonous Or Toxic Materials

Only those poisonous or toxic materials necessary for plant operation shall be present in the plant. Containers of poisonous or toxic materials shall be prominently labeled according to law for easy identification of contents and safely stored. Such materials shall be used only in accordance with label directions.

Satisfactory Compliance
This item will be satisfied when:

a. Poisonous or toxic materials not required for cleaning, sanitizing, insect or vermin control are not present in the plant.

b. Each of the following categories of poisonous or toxic substances shall be separated from each other:

   i. insecticides and rodenticides,

   ii. detergents, sanitizers, and related cleaning agents; and

   iii. caustic acids, polishes, and other chemicals.

c. Poisonous or toxic substances are not stored above shellfish or food-contact surfaces, except that this requirement does not prohibit conveniently locating detergents or sanitizers in utensil or equipment cleaning areas.

d. Poisonous or toxic materials are not used in a way that contaminates shellfish and food-contact surfaces nor in a way that constitutes a hazard to employees, and are used in full compliance with manufacturer's labeling.

Public Health Explanation
In order to reduce the potential for contamination, stored poisonous or toxic materials should be limited to those necessary to maintain the establishment. Proper labeling, use, storage, and handling are essential to prevent accidental contamination of shellfish and to ensure the safety of workers and the consumer.

12. Construction of Shucking Benches and Tables

Shucking benches and tables shall be designed and constructed so as to be easily cleaned. Contiguous walls, stalls, stands, and shucking blocks, if used, shall be similarly constructed.

Satisfactory Compliance
This item will be satisfied when:

a. Shucking benches and contiguous walls are constructed of easily cleanable, corrosion-resistant, impervious material and are free from cracks.

b. The tops of shucking benches and tables are located at an adequate height above the floor to prevent product contamination. Benches drain completely and rapidly, and drainage is directed away from any shellfish on the benches.

c. Shucking blocks are easily cleanable; are fabricated from safe material; are of solid, one-piece construction; and, unless an integral part of the bench, are easily removed from the shucking bench.

d. Stands or stalls and shuckers' stools, including padding, are fabricated from impervious, corrosion-resistant, safe materials and constructed so as to be easily cleaned and sanitized.

Public Health Explanation
Unless shucking benches, stands, blocks, and stalls are made of smooth material and are easily cleaned, they will become very dirty and may contaminate the shellfish.

13. Construction of Utensils and Equipment
Food-contact surfaces of utensils and equipment, including those used for handling ice shall be made of easily cleanable safe materials which will not easily disintegrate or crack. Utensils and equipment shall be so constructed as to be easily cleanable and shall be kept in good repair.

Satisfactory Compliance
This item will be satisfied when:

a. All utensils and equipment, including SSCA approved plasticware, are designed and fabricated from smooth, corrosion resistant safe materials, are durable under conditions of normal use, and are resistant to denting, buckling, pitting, chipping, and crazing.

b. There are no exposed screws, bolts, or rivet heads on food-contact surfaces, and all joints on food-contact surfaces are welded and have a smooth surface. If solder is used, it shall be composed of safe materials and be corrosion resistant.

c. Blower tanks, tubs, and skimmers, are constructed so that their top rims are at an adequate height above the floor to prevent product contamination.

d. Blower tanks, skimmers, returnable shipping containers, shucking buckets, and pans shall conform to the *Equipment Construction Guides*, Appendix A; except equipment in use prior to January 1, 1989; having seams soldered with safe materials which are corrosion resistant, smooth, and easily cleanable.

e. All utensils and equipment are in good repair.

f. All equipment, including external and internal blower air lines and hoses below a point 5.08 cm (two inches) above the overflow level of the tank and blower drain valves, are constructed so as to be easily cleanable; perforations in skimmers and colanders are smooth, to facilitate cleaning; all internal angles in the food-product zone are filled or otherwise fabricated to facilitate cleaning. The use of fiber mesh in the food-contact zone of equipment is not acceptable. However, the use of approved materials may be considered, provided it is properly designed and constructed for such usage and acceptable cleaning and sanitizing procedures are established.

Non-food-contact surfaces are constructed so they can be kept clean; seams and joints should be welded, whenever possible; outside seams should be welded or filled with solder; and there should be no inaccessible spaces in which dirt or organic material might accumulate.

g. Air-pump intakes are located in a protected place. Air filters are installed on all blower air-pump intakes. Oil bath type filters are prohibited.

Public Health Explanation
Colanders, shucking pails, skimmers, blowers, and other equipment or utensils which come into contact with the shucked shellfish and which have cracked, rough, or inaccessible surfaces or are easily cracked or chipped, or which are made of improper material, are apt to harbor accumulations of organic material in which bacteria or other microorganisms may grow. These microorganisms may later cause illness among those who eat shellfish, or spoilage in the shucked shellfish.

Slime and foreign material which accumulate in blower airpipes below the liquid level afford an excellent breeding place for bacteria. This material may be dislodged and forced into the batch of shucked shellfish in the blower, thus increasing the bacterial content of the shellfish.

14. General Maintenance and Cleanliness
Physical facilities of the plant shall be kept in good repair and cleaned as necessary to maintain sanitary conditions. Cleaning operations shall not cause contamination of food and food-contact surfaces. Detergents, sanitizers, and other supplies employed in the cleaning and sanitizing procedures shall be safe and effective for their intended uses. Miscellaneous and unused equipment and articles are not to be stored in rooms used for processing shellfish. No food or animals shall be permitted in the processing or storage area. Unauthorized personnel are to be excluded.

Satisfactory Compliance
This item will be satisfied when:

a. Building fixtures, walls, ceilings, and floors are cleaned at a frequency necessary to maintain an adequate level of sanitation. Frequency of cleaning will depend on such factors as type of process, condition of the plant, and hours of operation. Cleaning operations are conducted in such a manner as to minimize the danger of contamination of food and food-contact surfaces.

b. Detergents, sanitizers, and other supplies employed in the cleaning and sanitizing procedures are safe and effective for their uses.

c. Only material and equipment in routine use in the processing operation are stored in rooms used for shellstock storage, shucking, packing, repacking, or container storage. The premises are clean and free of litter and rubbish.

Public Health Explanation

The plant and building facilities should be kept clean so as to minimize the chance of contamination of shellfish during processing. Miscellaneous equipment and articles may interfere with plant operations and make clean-up more difficult.

15. Cleaning and Sanitizing Equipment and Utensils

All shucking benches and stools, shucking blocks, tables, skimmers, blowers, colanders, buckets, or any other equipment used in the processing operation shall be cleaned and sanitized as frequently as necessary during the day's operation to prevent the introduction of undesirable microbiological organisms and filth into the shellfish product. All food-contact surfaces of utensils and equipment shall be adequately cleaned and sanitized at the end of the day's operation and stored so as to protect against contamination. Cleaning and sanitizing may be required prior to commencing a day's operation. Refrigerators shall be kept clean.

Satisfactory Compliance
This item will be satisfied when:

a. Adequate cleaning facilities, including three-compartment sinks, brushes, detergents, sanitizers, hot water, and pressure hoses are available within the plant.

b. Cleaning compounds and sanitizing agents are free from undesirable microorganisms and safe and adequate under the conditions of use. Only chemical sanitizing agents at effective concentrations specified in 21 CFR 178.1010 (36) are used.

c. Procedures are established which may employ machines or devices used for cleaning and sanitizing equipment and utensils that will routinely render equipment and utensils clean and provide adequate sanitizing treatment.

d. Cleaning and sanitizing operations are conducted in such a manner and at such a frequency as is necessary to prevent the contamination of shellfish and food-contact surfaces.

e. Food-contact surfaces of utensils and equipment used in the plant are cleaned and sanitized prior to use and following any interruption during which food-contact surfaces may have become contaminated.

f. Cleaned and sanitized portable equipment and utensils are stored in such a location and manner that product surfaces are protected from splash, dust, and other contamination.

Public Health Explanation

Cleaned and sanitized equipment and utensils reduce the chance of contaminating shellfish during shucking and processing. Shellfish furnish an excellent growth medium for spoilage microorganisms, and small numbers of these microorganisms on improperly sanitized equipment may multiply to very high levels in the finished pack. (2, 5) Use of sanitizers are not effective unless the equipment is first thoroughly cleaned and rinsed.

Determining an adequate cleaning procedure for facilities and equipment will depend upon which method of sanitizing is selected and what equipment and utensils are identified to be washed in a sink or washed "in place." Detergents and brushes, including special brushes that may be needed for cleaning equipment, such as blower lines, should be available.

Cleaning and sanitizing of equipment and utensils should be initiated immediately after processing operations are finished. Postponing clean-up operations results in more difficult cleaning, creates conditions conducive to growth of bacteria and mold which may not be completely removed, and may result in product contamination.

16. Sources of Shellfish

Appropriate procedures shall be employed by the certified dealer receiving shellfish to assure that incoming shellstock are obtained either from a licensed harvester or a certified dealer, are properly tagged or otherwise identified to show their source, are accompanied by all required transaction records, and are clean and wholesome.

Satisfactory Compliance
This item will be satisfied when:

a. Shellstock originate from a harvester licensed according to Part I, Section F.1, an aquaculturist licensed according to Part I, Section G.1, or a certified dealer.

Transportation agents or common carriers utilized by certified dealers or licensed harvesters do not have to be certified.

b. Incoming shellstock are inspected to ensure they are clean and wholesome, alive, and are received at an appropriate temperature that will minimize microbial growth and retard decomposition as specified in Section B.2.

c. Shellstock are identified in accordance with Section B.5 and are protected from contamination.

d. Unwholesome, unidentified, dead, or contaminated shellstock are rejected or discarded.

Public Health Explanation

Certified dealers are responsible to assure that shellfish purchased for direct sales, further shipments, or processing are safe and wholesome. The safety of shellfish are predicated on the cleanliness of the growing area waters from which they are obtained, and the sanitary practices applied during harvesting and shipping.

The positive relationship between sewage-polluted shellfish and enteric disease has been demonstrated many times (37, 38, 39, 40, 41). Because physiologically active shellfish pump and filter large quantities of water as part of their feeding process, rapid intake and concentration of bacteria, viruses, marine toxins, and other poisonous and deleterious substances may occur (18, 42, 43, 44, 45, 46, 47, 48, 49). Therefore, the shellfish may contain higher levels of chemical contaminants or pathogens than are found in the water in which they grow.

The shellfish-water bacteria ratio depends upon the shellfish species, water temperature, presence of certain chemicals, and varying physiological capabilities of the individual animals (42, 43, 44). If the water in which the shellfish are grown is polluted, it may be assumed that the shellfish will also contain pathogenic bacteria or viruses capable of causing disease in man.

Furthermore, there is evidence that some pathogenic organisms will survive in shellfish for a considerable length of time after harvesting and that some bacterial pathogens may multiply in the absence of adequate refrigeration (18, 27, 50). In addition, shellfish contaminated by added trace metals can result in illness to man if consumed in sufficient quantities (51, 52). Health hazards also may result from the presence of naturally occurring biotoxins produced by certain marine dinoflagellates (53, 54, 55, 56). The occurrence of these poisons is related to the concentration of toxic dinoflagellates in the growing area. The contamination of shellfish by these dinoflagellates usually occurs in well-defined areas and, in some instances, only during certain seasons (54, 55) not widespread over all shellfish producing areas.

Cooking does not necessarily ensure safety of contaminated shellfish since, in ordinary cooking processes, shellfish may not be heated sufficiently to ensure a kill of pathogenic organisms, although a considerable reduction will take place (19, 33, 57). Also, normal cooking processes cannot be relied upon to destroy paralytic shellfish poison (54).

Certified dealers have three principal responsibilities to assure that the consumer receives a safe product. The first is to purchase only safe and wholesome raw products. The second is to maintain the product in a sanitary manner. The final responsibility is to ship the product under sanitary conditions. The tagging and shipping records requirements, the sanitary shipping practices requirements, and the raw product inspection requirements are necessary to fulfill these responsibilities.

17. Shucking

Shellfish shall be shucked in a manner such that they will not be subjected to contamination. Shellstock shall be reasonably free of mud when shucked. Only wholesome and safe shellfish shall be shucked, and shellfish with badly broken shells shall be discarded. Shucking operations should be scheduled to ensure that shucked product does not remain at the shucking station for prolonged periods, and to minimize commingling of shellfish from different sources. Water used in fluming or washing shellstock shall be from an approved source.

Satisfactory Compliance
This item will be satisfied when:

a. Shellfish are not subject to contamination while being held or processed. This shall be accomplished when:

> i. shellstock to be shucked are stored in a sufficient height off the floor or in such locations that contamination from standing water or splash from foot traffic does not occur.

> ii. shellstock are reasonably free of mud when shucked;

> iii. only safe and wholesome shellfish are shucked; and

> iv. dead shellfish or shellstock with badly broken shells are discarded; and

> v. other manufacturing operations which could result in contamination of shellfish are separated by adequate barriers or space.

b. Shellfish are not subject to contamination or held for an excessive time at unsafe temperatures during shucking and packing. This shall be accomplished when:

> i. shucking buckets and storage containers are so used that their rim is at an adequate height above the floor to prevent contamination from floor splash;

ii. shucked meats shall be delivered to the packing room within one (1) hour;

iii. shucking buckets are completely emptied at the packing room and no overage is returned to the shucker;

iv. shucking containers are rinsed clean with running water and sanitized before each filling; and

v. the precautions that apply to hand shucking methods are applied to mechanical procedures for the shucking of all species of shellfish.

c. The use of "dip" buckets for hand or knife rinsing is prohibited.

d. Shellstock from different lots are kept separate. If a SSCA permits commingling, it shall develop a commingling management plan. The objective of any state management plan is to minimize the commingling dates of harvest and harvest areas. The plan shall limit the practice of commingling to primary dealers and to shellstock purchased directly from harvesters from specific areas as defined by the SSCA. The plan shall further define how the commingled shellfish will be identified. The SSCA will define how the commingled shellfish will be identified. The SSCA will define the primary level dealer. Shucking operations are scheduled to avoid commingling shellfish from different lots.

e. Water used for fluming or washing shellstock and shucked product is obtained from an approved source.

Public Health Explanation

If shellfish are not reasonably clean at the time of shucking, a considerable quantity of the adhering material will be mixed into the shucked shellfish during the shucking process, thus contributing to high bacteria counts in the final product (29).

(See Public health explanation, Section B.3, Washing of Shellstock). Although the primary responsibility for washing shellstock rests with the harvester, the certified dealer is not relieved of responsibility for compliance with this item.

The bacteria count of the final pack also is related to the elapsed time after shucking when the shellfish are held at temperatures favorable to the rapid growth of bacteria (13). Factors which influence the length of time required to lower the temperature of shucked shellfish to 7.2°C (45°F) include the temperature of blower or other process water, the speed of the individual shucker or shucking machinery, the frequency with which the shucking containers are delivered to the packing room, ambient air temperature in the plant, and the temperature of the shellstock being shucked (13, 14, 17, 19). To maintain optimum bacteriological quality, it is preferable that the elapsed time between shucking and cooling to a temperature of 7.2°C (45°F) does not exceed four hours. More rapid processing is very desirable (17).

Return of overage from the packing room to the shucker and bench grading of shellfish are also practices which may result in at least a portion of the shellfish being held on the shucking bench for prolonged periods and permit undesirable microbial growth. It is especially important that all grades be delivered to the packing room in a timely manner when temperature of the meats exceeds 7.2°C (45°F). To encourage frequent delivery of the shucked shellfish to the packing room, the shucking containers may be limited to a size that an average shucker might reasonably be expected to shuck full in one hour. The quantity of shellstock at the shucking station may similarly be limited. Storage of shucked shellfish on the shucking benches for long periods of time also increases the possibility of contamination of the shucked shellfish by splash or flies.

Bacteriological examination of the water in dip buckets has shown very high coliform counts. Since water from the dip bucket may be carried out into the shucked shellfish, there is need to control the sanitary quality of the water.

18. Shell and Waste Disposal
Shells from which meats have been removed, and other non-edible materials shall be removed promptly from the shucking room and disposed of so as not to create nuisance conditions.

Satisfactory Compliance
This item will be satisfied when:

a. Shells and other non-edible materials are promptly and effectively removed from the shucking bench or table area to prevent interference with the sanitary operations of the shucking process.

b. Shell and waste materials are disposed of so as to minimize the development of odor and not become an attractant, harborage, or breeding place for vermin. Fly control measures may be necessary in the vicinity of shell piles. In some instances, such as with surf clams, culled shellfish and unused portions of body meats may need to be separated from empty shells in order to prevent insanitary or nuisance conditions from developing at the disposal.

c. Shells and non-edible waste are disposed of in such a manner that contamination of food, food-contact surfaces, ground surfaces, and water supplies does not occur.

Public Health Explanation
Shellstock shipping and shucking facilities can protect against infestation by vermin if building entrances are protected, the grounds do not provide harborage, and there is no food available in the buildings or on the grounds. Removing shell and organic processing wastes from the plant and properly disposing of these wastes can play a key role in controlling vermin. Methods found to be suitable for removing these materials without contaminating the shucked product include conveyors, baskets, barrels, wheelbarrows, and shell drop-holes.

When shells are to be temporarily piled or stored on the premises, special controls may be needed. Organic wastes, including culled shellfish, clam siphons, and surf and ocean quahog viscera, need to be discarded into separate containers from the shells in the plant during shucking. These wastes can then be disposed of separately from the shell at, for example, a landfill.

Fly control measures, such as insecticide spraying, may also be necessary on the shell pile.

Proper disposal and prompt removal of shell and non-edible wastes from the plant also makes it possible to keep the premises clean, and decreases the likelihood that any product or food-contact surfaces will become contaminated.

19. Construction and Handling of Single-Service Containers

All single-service and single-use containers shall be fabricated from safe materials and so designed to be easily cleaned and sanitized. Containers shall be stored and handled in a sanitary manner and, where necessary, shall be cleaned and sanitized immediately prior to filling.

Satisfactory Compliance
This item will be satisfied when:

a. Containers for shucked shellfish are clean; constructed of non-toxic metal, food-grade plastic, glass, or other impervious material; and designed and fabricated such that the contents will be protected from contamination during shipping and storage. Covers of returnable containers are designed so as to protect the pouring lip of the container from contamination.

b. Single-service and single-use containers and covers are kept in original cartons until used, are kept clean and dry, or are otherwise protected. Container-storage rooms are kept clean and free of vermin; containers are stored in a manner that the presence of vermin may be easily detected; and container-storage rooms are not used as general store rooms for unused equipment and materials. Containers are stored off the floor and away from the walls to facilitate inspecting and cleaning the area.

c. Containers which may have become contaminated or unclean during storage are cleaned, sanitized, and properly stored prior to filling, or are discarded.

d. Plant employees use reasonable precautions to prevent food-contact surfaces of containers from coming into contact with themselves or their clothing. Containers in the packing rooms are protected and kept inverted on stands or tables at adequate height above the floor to prevent contamination from splash.

Public Health Explanation
Single-service and single-use containers which have not been stored and handled in a sanitary manner may become contaminated and thus may contaminate the packaged shellfish.

20. Packing of Shucked Shellfish

Shucked shellfish shall be promptly packed without being exposed to contamination. Shucked shellfish shall be packed and shipped in clean containers fabricated from safe materials. Returnable containers shall be accepted only for interplant shipment of shucked shellfish and shall be sealed during transport.

Satisfactory Compliance
This item will be satisfied when:

a. Skimmer tables and other packing equipment are located so they will not receive drainage from the delivery window or contamination from shucking room equipment and utensils.

b. Shuckers and other unauthorized persons do not enter the packing room for any purpose. An exception may be made in a small operation where an employee may work in both the packing room and shucking room. In such cases, the employee shall put on a clean apron or other cover before entering the packing room and wash his hands thoroughly after entering.

c. Shellfish meats are examined for naturally occurring extraneous materials such as shell fragments, sand, pearls, and other non-edible components. Also, packaging processes and equipment do not transmit contaminants or objectionable substances to the products, conform to applicable food additive regulations, and provide protection from contamination or adulteration.

d. Shucked meats are thoroughly drained, cleaned as necessary, and packed promptly after delivery to the packing room. The packing operations are scheduled and conducted so as to chill all meats to an internal temperature of 7.2°C (45°F) within two (2) hours of delivery to the packing room. Containers of shucked shellfish are closed promptly after filling. Shucked meats which are to be packed into containers having a capacity of more than 3785 ml (one gallon - 128 ounces) shall be pre-chilled to 7.2°C (45°F) or less prior to packing.

e. Washing, blowing, and rinsing of shellfish meats are in compliance with the time limits specified in 21 CFR 161.130 (see Appendix B).

f. Shucked shellfish are packed only into containers labeled with the authorized plant certification number. Storage of usable containers or covers with a certification number other than that on the unexpired plant certificate will be considered a violation of this item. It is recommended that containers be sealed in such a manner that tampering can be detected.

g. Returnable containers are used only for interplant shipments of shucked shellfish and are sealed during transport. These shellfish shall be repacked into proper single-service or single-use containers.

Public Health Explanation

The packing step in the shucking and packing of shellfish is a critical control point. The shucked meats can provide a growth substrate for spoilage and pathogenic bacteria if proper sanitation and temperature controls are not applied. As such, shellfish are considered a potentially hazardous food. Controls that should be applied include using clean packing equipment, using packers with clean hands and outer garments, washing and packing the shellfish in a timely manner, using clean shipping containers, and promptly refrigerating the shellfish. Prompt packing and refrigeration is particularly important in order to prevent the shucked meats from being in excess of 7.2°C (45°F) for more than two (2) hours after delivery to the packing room.*

Unacceptable practices that can interfere with the prompt handling, packing, and refrigerating of shellfish include holding shucked meats at the shucking station for prolonged periods, return of overage to the shucker, and bench grading of shucked meat. Anther frequently encountered unacceptable practice is soaking of shucked meats for prolonged periods in water for the purpose of increasing yield through uptake of fresh water by the shellfish. The standards of identity for oysters is set forth in the Code of Federal Regulations, 21 CFR 161.130 (Appendix B).

Similar requirements are recommended for other shellfish species.

Shellfish meats in unchilled water and/or packed at temperatures above 10°C (50°F) in 3785 ml (one-gallon) containers cannot be cooled to 7.2°C (45°F) in the required time. Therefore shellfish meats intended for large capacity containers must be prechilled. The graphs of cooling rates in Appendix C which were based on a study by the American Can Company show that it takes approximately two (2) hours for oysters in one-gallon metal containers packed in wet ice to cool from 10°C (50°F) to 7.2°C (45°F). Cooling rates in plastic containers would be even longer. Chilled water or flaked ice in the blower, at the skimmer, or in chilling containers are needed for rapid chilling. The ice should be melted and drained prior to final packing in the shipping container. Another option would be packing shucked meats in small consumer-size containers which can be chilled more rapidly than large sizes.

21. Labeling Shucked Shellfish

Each individual package of fresh or fresh frozen shucked shellfish shall have permanently recorded on the principal display panel all information required by 21 CFR 101 and 21 CFR 161.130-161.140 (36) and the certification number of the dealer. Additionally, each individual package of fresh or fresh frozen shucked shellfish with a capacity of less than 1873 ml (one-half gallon - 64 ounces) shall have a SELL BY date; each individual package with a capacity of 1873 ml (one-half gallon - 64 ounces) or more shall have the DATE SHUCKED. Packages of shucked shellfish containing 1873 ml (one-half gallon - 64 ounces) or more shall have the

---

*Internal temperature is measured by reading the thermometer with the bulb inserted into the approximate geometric center of the container.

certification number and name of the packer on the side wall of the package and the date marking on both the lid and the side wall or bottom. For packages of shucked shellfish containing 1873 ml (one-half gallon - 64 ounces) or more, the side wall is considered the principal display panel since the cover may not remain an integral part of the package. Frozen shellfish shall be clearly labeled as such.

Satisfactory Compliance
This item will be satisfied when:

a. Labeling of all packages of fresh or fresh frozen shucked shellfish is in conformity with the requirements of 21 CFR 101 for information on the principal display panel and with 21 CFR 161.130-161.140 for standard of identity. (Appendix B contains the applicable requirements from the CFR).

b. The principal display panel on each package of fresh or frozen shucked shellfish with a capacity of less than 1873 ml (one-half gallon - 64 ounces) contains the certification number of the packer and the words SELL BY followed by a recommended last date of sale of the product which provides a reasonable subsequent shelf-life. The date will consist of the common abbreviation for the month and number of the day of the month. For frozen shellfish, the year will be added to the date.

c. For each package of fresh or frozen shucked shellfish with a capacity of 1873 ml (one-half gallon - 64 ounces) or more, the word SHUCKED followed by the actual shucking date shall appear on the lid and on the side wall or bottom of durable containers. The date shall consist of the common abbreviation for the month and number of the day of the month or the number of the day of the year (Julian calendar day). For frozen shellfish, the year will be added to the date.

d. Frozen shellfish are labeled as frozen in type of equal prominence immediately adjacent to the name of the shellfish.

e. All required information is provided in a legible and indelible form.

f. The labeling requirements of this section do not apply to returnable containers used to transport shucked shellfish between certified dealers for the purpose of further processing or repacking. When this operation is practiced, the shipments are accompanied by a transaction record containing the original shucker's certification number and name, shucking date, the quantity of shellfish per container and the total number of containers. Labeling requirements also do not apply to master shipping cartons as long as the individual containers within the carton are properly labeled.

Public Health Explanation

The Federal Food, Drug and Cosmetic Act requires that food labels provide an accurate statement which includes the name and address of either the manufacturer, packer, or distributor; the net amount of food in the package; the common or usual name of the food; and the ingredients, unless the product conforms to standard of identity requirements. Foods shipped in interstate commerce having labels that do not meet these requirements are deemed misbranded and in violation of Section 405 of the Food, Drug and Cosmetic Act.

The NSSP further requires that the product be identified with certain information showing that the shellfish were harvested by licensed diggers and shipped and processed by certified dealers. This information assists in tracing the product back through the distribution system to the harvest area in the event the shellfish are associated with a disease outbreak. The requirement for placing the certificate number and date marking on the side wall or bottom of durable containers holding 1873 ml (64 fluid ounces) or more is to discourage re-use of these containers for illegal purposes. The ISSC has voted to prohibit use of rubber stamps or adhesive labels as a deterrent to bootlegging.

22. Refrigeration and Shipping of Shucked Shellfish

After shucking and packing in accordance with Section D.20, shucked shellfish are adequately refrigerated and protected to prevent contamination and minimize product deterioration.

Satisfactory Compliance
This item will be satisfied when:

a. Shucked shellfish are held and transported at temperatures of 7.2°C (45°F) or less. Storage and shipping of sealed containers of shucked shellfish in wet ice is highly recommended. Shucked shellfish shall also be shipped in compliance with the provisions of Section B.2.

b. Packaged shellfish to be frozen are arranged to ensure rapid freezing, and are frozen at an ambient temperature of -17.8°C (0°F) or less, with packages frozen solid within 12 hours after the start of freezing. Frozen shellfish shall be handled in such a manner as to remain frozen solid, and are held at -17.8°C (0°F) or less.

c. Refrigeration and frozen-storage compartments are equipped in compliance with Section D.2.f.

d. All containers holding shucked shellfish are kept covered during refrigeration.

Public Health Explanation

Shucked shellfish are an excellent medium for the growth of bacteria. Thus, it is very important that the packaged shellfish be cooled and refrigerated promptly so that bacteria growth is minimized. Studies have shown that bacterial growth is significantly reduced at storage temperatures of less than 7.2°C (45°F) and that storage in wet ice is the most effective method for refrigeration of shucked meats (13, 14, 17). Alternate freezing and thawing of shellfish should be avoided in order to retard deterioration and spoilage.

23. Ice

Ice shall be made in a sanitary manner or obtained from a safe source specifically approved by the appropriate state regulatory agency. Ice shall be stored and handled in a sanitary manner.

Satisfactory Compliance
This item will be satisfied when:

a. Ice is manufactured at the establishment from potable water in a commercial machine which has been properly installed without any cross connections, or in another establishment approved by the appropriate regulatory agency.

b. Ice is stored so as not to come into contact with non-clean surfaces and is handled in such a manner that it will not be contaminated. Equipment used to handle ice is kept clean and stored in a sanitary manner.

c. Ice not manufactured in the shellfish processing establishment is inspected upon receipt and rejected if not delivered in clean conveyances and protected from contamination.

Public Health Explanation
Ice may be contaminated by non-potable water or may become contaminated during freezing or in subsequent storing and handling. When non-hermetically sealed containers of shellfish are stored in unsanitary ice, a partial vacuum may form within the containers and draw water from the melting ice into the container and contaminate the packed shellfish. Special attention should be given to ice used for direct contact chilling of shellfish meats to assure that the ice is of acceptable quality.

24. Records

Complete and accurate, legible transaction records shall be maintained by each certified dealer which provides all information necessary to trace all purchases and sales of shellfish back to their source.

Satisfactory Compliance
This item will be satisfied when:

a. Complete, accurate and legible records must be maintained by each certified dealer. The necessary information and the required standard format for maintaining the information is illustrated in Appendix D. These records are sufficient to document that the shellfish are from an approved source and to permit a container of shellfish to be traced back to its specific incoming lot. Purchases and sales are recorded in a permanently bound ledger book and are maintained for a minimum of one (1) year. Transaction records indicating origin, date, and time of receipt of the product are maintained in a legible, orderly file. If records are maintained on a computer, they must contain all the necessary information in Appendix D.

b. Records covering purchases and sales of fresh shellfish are retained for a minimum of one year. Records covering purchases and sales of frozen shellfish should be retained for at least two years or for a period of time that exceeds the shelf-life of the product, if that is longer than two (2) years.

Public Health Explanation

In case of an outbreak of disease attributable to shellfish, it is necessary that health departments and other appropriate state and federal agencies be able to determine the source of contamination, and thereby to prevent any further outbreaks from this source. This can be done most effectively by following the course of a shipment, through all the various dealers who have handled it, back to the point of origin by means of records kept by the shellfish dealers.

Maintaining adequate records are considered by some industry members to be a burden. This has resulted in various unacceptable practices being encountered by health officials, including no written records of purchase, undated shippers tags maintained in an unordered manner, new shipping tags being placed on a lot without records to correlate the original identity of the lot with the new identity, and shellfish on the premises with no tags. Although these dealers often have "records" in the most general sense, these records are not in the form which meets the intent of the NSSP certification requirement to provide traceability on a lot-by-lot basis. As a result, follow-up investigations of disease outbreaks have been stymied, identification of the cause of the outbreak has been delayed, and outbreaks have continued. Appendix D contains an example of a typical ledger that may be used to provide the required information.

An example where the failure to maintain adequate records was identified as one of the principal contributing factors to a series of continuing disease outbreaks was in 1981 and 1982. The outbreaks continued for several months and affected thousands of people. An investigation by the states involved and FDA revealed that some states were unable to enforce the recordkeeping and tagging requirements of the NSSP. FDA found in one state that approximately one-third or the certified dealers inspected failed to maintain adequate records. State officials realized that an improved labeling or manifest system was needed to track shellfish in the marketplace back to the distributor and to the digger. However, they also recognized that no single source identity and recordkeeping system will be applicable to all situations in each state. Therefore, specific requirements should be developed by each state to achieve the NSSP requirements.

25. Employee Health

Any person infected with a disease in a communicable stage or while a carrier of such disease or who has an infected wound or open lesion on their body, or other abnormal sources of microbiological contamination, shall be excluded from the shucking or packing plant. A person-in-charge who has reason to suspect that any employee has contracted a communicable disease shall immediately notify the proper health officials. Pending appropriate action by health officials, the suspect employee shall be excluded from the plant.

Satisfactory Compliance
This item will be satisfied when:

a. Persons infected by disease in a communicable form, or while a carrier of such disease, or while infected with boils, sores, infected wounds, or acute respiratory infection shall not work in a shellfish processing establishment in any capacity in which there is a likelihood of such persons contaminating shellfish or shellfish-contact surfaces with pathogenic organisms or transmitting disease to other persons.

b. Daily observation of employees are made by the supervisor, with reasonable inquiries being made when signs of illness appear. Employees having diarrhea, sore throat or any other symptoms of illness or disease promptly report this to their supervisor.

c. Upon an inquiry indicating the possibility of a communicable disease, the infected employee is excluded from the plant pending clearance by a licensed medical doctor.

Public Health Explanation

It is considered good public health practice for any person who, by medical examination or supervisory observation, is shown to have, or appears to have, an illness, open lesion, including boils, sores, or infected wounds, or any other abnormal source of microbial contamination by which there is a reasonable possibility of food, food-contact surfaces, or food-packaging materials becoming contaminated, to be excluded from any operations which may be expected to result in such contamination until the condition is corrected. Personnel should be instructed to report such health conditions to their supervisors.

26. Supervision

The management shall clearly designate a competent individual to be accountable for compliance with the items of this manual relating to personal hygiene and plant sanitation.

Satisfactory Compliance
This item will be satisfied when:

a. A reliable, competent individual has been designated by the management to supervise general plant operations as enumerated in this section. Designating such an individual does not relieve management of the responsibility for complying with these items.

b. There is evidence that supervisors have been monitoring employee hygiene practices, including handwashing, eating and smoking at work stations, and storing personal items of clothing. Supervisors shall also ensure that proper sanitary practices are implemented, including plant and equipment clean-up, protecting shellfish from contamination during shucking and packing and rapid product handling.

c. Unauthorized persons are not permitted in the processing areas during periods of operations.

d. No animals are permitted in the processing area.

Public Health Explanation

Handwashing by employees is an important public health measure. Unless someone is made specifically responsible for this practice, it is apt to be forgotten or overlooked. Similarly, one person must be responsible for plant clean-up. In general, it is considered to be good practice to clearly assign supervisory personnel the responsibility for assuring compliance by all personnel with requirements of this section.

27. Personal Cleanliness

All persons working in direct contact with shellfish processing operations or food-contact surfaces shall maintain a high level of cleanliness and personal hygiene.

Satisfactory Compliance
This item will be satisfied when:

a. Employees handling shucked shellfish wear clean outer garments. These outer garments are rinsed or changed as necessary to be kept clean.

b. Employees wash their hands thoroughly with soap and water and sanitize their hands in an adequate handwashing facility before starting work, after each absence from the work station, after each interruption and at any other time when their hands may have become soiled or contaminated. Utensil sinks are not used for handwashing.

c. Finger cots, gloves, and shields, if worn by shuckers, are sanitized as often as necessary or at least twice daily; are properly stored until used (see Section D.20); and are maintained in an intact, clean, and sanitary condition. Finger cots, gloves, and shields should be made of an impermeable material except where use of such materials would be inappropriate or incompatible with the work involved.

d. Hands of employees handling shucked shellfish are either protected by sanitized finger cots or gloves, or the hands are washed and disinfected *immediately* before any manual handling of the shucked shellfish.

e. Employees do not store clothing or other personal belongings, eat food or drink beverages, use tobacco in any form or spit in areas where shellfish are shucked or packed or in areas used for washing equipment or utensils.

f. Employees handling shucked shellfish wear effective hair restraints, remove all insecure jewelry and remove from hands any jewelry that cannot be adequately sanitized.

g. Employees take other necessary precautions to prevent contamination of shucked shellfish with microorganisms or foreign substances, including, but not limited to, perspiration, hair, cosmetics, chemicals, and medicants.

Public Health Explanation

The hands of all employees frequently come into contact with their clothes. Hence, it is important that the clothes worn during the handling of shucked shellfish be clean. The nature of the work makes it necessary that protective outer garments be worn. Finger cots, gloves and shields, unless effectively sanitized periodically, will accumulate bacteria which may contaminate the shucked shellfish. Disease-producing agents may be carried on the hands of shuckers and packers unless proper handwashing is practiced. Employees handling shucked shellfish need to sanitize their hands as an added public health control practice. A container of effective bactericidal solution should be present in the packing room during periods of operation.

28. Education and Training

Employees handling shucked shellfish should receive appropriate training in proper food-handling techniques and should understand the danger from poor personal hygiene and insanitary practices.

Satisfactory Compliance
This item will be satisfied when:

a. Supervisors have received appropriate training in proper food-handling techniques and food protection principles and are cognizant of personal hygiene and sanitary practices.

b. Employees receive instruction and training in proper food handling and personal hygiene and sanitary practices from supervisory personnel or from other sources acceptable to the state.

c. Insanitary practices are brought to the employees attention by supervisors and the employees are instructed on the proper sanitary practice that is to be used.

Public Health Explanation
Employees engaged in shucking and packing operations may not fully understand the principles of good personal hygiene and sanitary food-handling practices, or may not fully appreciate the importance of following these practices. Management has the responsibility to train employees to follow good personal hygiene and sanitary practices, stress the importance that the practices be followed, set an example in their implementation, and correct situations where proper practices are not followed.

Personnel responsible for identifying sanitation failures or food contamination should have a background of education or experience, or a combination thereof, to provide a level competency necessary for production of clean and safe shellfish. Food handlers and supervisors should receive appropriate training in proper food handling techniques and food protection principles and should be informed of the danger of poor personal hygiene and insanitary practices.

(This page is blank.)

Section E

Shellstock Shipping

A shellstock shipper may buy and sell shellstock from a harvester or other certified dealer, may reship shellstock or shucked shellfish, and may relabel shellstock. Repackaging may only be done by shellstock shippers with permanent physical facilities. A shellstock shipper may not shuck, relabel, or repack shucked shellfish.

The sanitation requirements which apply to a shellstock shipper vary depending upon the operations in which the dealer is engaged. The most stringent requirements apply to dealers who repackage or relabel since extra recordkeeping requirements apply to these operations. Firms certified as shucker-packers and repackers may also ship shellstock under their shucker-packer (SP) or repacker (RP) certification number.

1. Source, Identification and Records
All shellstock shall originate from an approved source and be packaged, protected and identified according to the requirements of Sections B and D.

Satisfactory Compliance
This item will be satisfied when:

a. All incoming shellstock are inspected to assure compliance with requirements of Section D.16.

b. Dead, unwholesome, inadequately protected, or unidentified shellfish are rejected or discarded.

c. Complete, accurate, and legible records must be maintained by each certified dealer. The necessary information and the required standard format for maintaining the information is illustrated in Appendix D. These records are sufficient to document that the shellstock are from an approved source and to permit a package or bulk shipment to be traced back to the harvest area, date of harvest, and the harvester. Purchases and sales are recorded in a permanently bound ledger book and are maintained for a minimum of one (1) year. Transaction records indicating origin, date, and time of receipt of the product are maintained in a legible, orderly file. If records are maintained on a computer, they must contain all the necessary information in Appendix D.

Public Health Explanation
It is important that all shellstock be obtained from certified shippers to diminish the possibility of receiving contaminated shellfish. Shellstock should be received in good condition and be held

and repacked under such conditions and temperatures to minimize deterioration and contamination. Excessively high temperatures are known to produce undesirable increases in bacterial levels and may lead to significant increases in environmental pathogens present in some species of shellfish (27, 33). Tagging and recordkeeping are important in the event a question arises about the quality or safety of a product and it is necessary to trace the shellfish back to its source.

2. Shellstock Storage and Shipping

Shellstock shall be shipped and stored at such temperatures and under such conditions as are necessary to minimize the potential for microbial growth and product deterioration, and to prevent contamination. Shellstock shall be identified and records maintained in such a manner that containers can be traced back to their source.

Satisfactory Compliance
This item will be satisfied when:

a. Trucks used to store or transport shellstock are constructed and maintained, and cleaned according to the requirements of Section B.1.d, f, g, h, and i. Shellstock are transported in adequately refrigerated trucks when the shellstock have been previously refrigerated or when ambient temperatures are such that unacceptable bacterial growth or deterioration may occur (see Section B.2).

b. Buildings in which shellstock are stored or repacked comply with the requirements of Sections D.1 through D.11 and D.14. Shellstock shippers who store or repack shall have access to sanitary toilet facilities acceptable to the SSCA and an approved water supply providing at least warm water suitable for hand washing.

c. Shellstock in storage are protected from contamination and maintained at temperatures necessary to minimize microbial growth pursuant to the requirements of Section D.2.

d. All equipment and conveyances which come into contact with shellstock are maintained and cleaned according to the requirements of D.14.a.

e. Ice used for shellstock refrigeration is manufactured, stored, and handled in accordance with Section D.23.

f. Shellstock are identified in accordance with the requirements of Section B.5.e and records maintained in accordance with the requirement of Section D.24. Shippers whose physical facilities consists of trucks and/or docking facilities only, shall not repackage shellfish and shall have a business address at which records are maintained and inspections can be performed. Shellstock shippers who harvest and store or repack shellstock shall either have their own facility for proper storage of shellstock or have arrangements with a facility approved by the SSCA for storage of shellstock.

g. Shellstock from different sources are separated to prevent commingling. If a SSCA permits commingling, it shall develop a commingling management plan. The objective of any state management plan is to minimize the commingling dates of harvest and harvest areas. The plan shall limit the practice of commingling to primary dealers and to shellstock purchased directly from harvesters from specific areas as defined by the SSCA. The plan shall further define how the commingled shellfish will be identified. The SSCA will define the primary level dealer.

h. Supervision is provided in accordance with the requirements in Section D.26.

Public Health Explanation
The sanitary requirements for individual shellstock shippers are highly variable since they may engage in several different phases of processing and distribution. Some shellstock shippers may have only a truck which is used to ship shellstock from the harvester to a processor or the market. Other shippers must have a building where shellstock is stored, repacked, or relabeled. Consequently, the applicable sanitary controls must be based on an evaluation of the individual characteristics of the operation.

3. Repacking and Relabeling Shellstock
Only clean and wholesome shellfish shall be repacked. Repacking facilities and equipment shall meet applicable sanitation requirements to assure that the shellfish are not contaminated during repacking and microbiological deterioration does not occur. Shellstock from different lots shall not be commingled. Each container of repacked or relabeled shellstock shall be identified as to harvest area, date of harvest, type and quantity of shellfish, and the certification number of the shellstock shipper. Records shall be maintained which will permit a package of shellstock to be traced back to the harvest area. Records shall also include the date of harvest and, if possible, the harvester or group of harvesters.

Satisfactory Compliance
This item will be satisfied when:

a. Only shellstock that are clean, alive, and wholesome are repacked or relabeled.

b. Shellstock repacking facilities are in compliance with requirements of Section E.2.b.

c. Shellstock from different lots shall not be commingled during repacking.

d. Sacks, boxes, and other shellstock packing containers are clean and fabricated from safe materials.

e. Each certified firm affixes an approved, durable, waterproof tag of minimal size - 6.7 by 13.3 cm (2 5/8 by 5 1/4 inches) - to each container of shellstock prior to shipment.

f. The dealer's tags shall contain legible information arranged in the specific order as follows:

    i. the dealer's name, address, and the certification number assigned by the SSCA;

    ii. the original shipper's certification number including the state abbreviation;

    iii. the date of harvesting;

    iv. the most precise identification of the harvest location or aquaculture site as is practicable (e.g. Long Bay, Smith's Bay, or a lease number);

    v. type and quantity of shellfish; and

    vi. the following statement will appear in bold capitalized type "THIS TAG IS REQUIRED TO BE ATTACHED UNTIL CONTAINER IS EMPTY AND THEREAFTER KEPT ON FILE FOR 90 DAYS."

Public Health Explanation

The shellstock shipper plays an important role in maintaining the chain of controls upon which the NSSP certification program is based. Proper identification of repackaged or relabeled shellfish and adequate records are emphasized because investigation of past shellfish-related disease outbreaks have revealed on numerous occasions that commingling of shellstock from different lots and incomplete or inadequate labeling and transaction records have prevented investigators from determining the source of the shellfish. The other controls focus on protecting shellstock from being contaminated during repacking and storage and on maintaining the shellstock at temperatures which minimize bacterial growth and product deterioration.

Section F

Repacking

Repacking is the process of removing shucked shellfish from one package and placing them in another. Repacking of shucked shellfish exposes them to additional handling and increases the possibility of contamination. Combining shellfish from more than one growing area or dealer into one package diminishes the state shellfish control agency's ability to trace shellfish back to the source if questions of the product's safety arise, and permits the possibility of the whole package becoming contaminated if any one source is contaminated. For these reasons, repacking should be discouraged and every effort made to package shellfish in the final container at the time of the initial packing.

1. Repacking of Shucked Shellfish

Shucked shellfish to be repacked shall originate only from a certified shucker-packer and, upon receipt, shall be refrigerated, protected and labeled in compliance with this Manual. Records of each purchase shall be maintained by the dealer which will permit all shucked shellfish to be traced back to the source. Shellfish from different lots shall not be commingled during repacking. The internal temperature of the fresh shellfish shall be 7.2°C (45°F) or less than −17.8°C (0°F) for frozen shellfish at the time of receipt. Only wholesome shellfish shall be repacked and applicable Manual requirements shall be followed to minimize microbial growth and product deterioration.

Satisfactory Compliance
This item will be satisfied when:

a. Facilities in which shucked shellfish are repacked are in compliance with Sections D.1. and D.3.-11.

b. Shucked shellfish for repacking originate only from certified shucker-packers, are labeled in compliance with Section D.21, and are received at temperatures of 7.2°C (45°F) or less.

c. Complete, accurate, and legible records must be maintained by each repacker. The necessary information and the required standard format for maintaining the information is illustrated in Appendix D. These records will permit a package of repacked shellfish to be traced back to the original shucker-packer. The records will include date of pack and source of the shellfish. Purchases and sales are recorded in a permanently bound ledger book. Records are maintained for a minimum of one (1) year for shucked shellfish and two (2) years for frozen shellfish. If records are maintained on a computer, they must contain all the necessary information in Appendix D.

d. Shucked fresh shellfish are maintained at an internal temperature of 7.2°C (45°F) or less while in storage and throughout repacking operations. Repacking at less than 1.7°C (35°F) and storage in wet ice is recommended. Frozen shellfish which are thawed for repacking do not exceed 7.2°C (45°F) in any portion of thawed shellfish.

e. Only wholesome shellfish are repacked and the requirements of Sections D.13.-20. are followed to prevent contamination and to minimize microbial growth and product deterioration.

f. Repacked shellfish are labeled in compliance with Section D.21. and the original date of shucking shall be considered in establishing the SELL BY date.

g. Ice is manufactured and handled in accordance with the requirements of Section D.23.

h. Employee health, supervision, and education and training are in compliance with Sections D.25-28.

i. Shucked shellfish from different lots shall not be commingled during repacking.

Public Health Explanation

Repacking of shucked shellfish exposes shucked meats to additional handling and increases the possibility of introducing contamination or subjecting them to increased temperatures which may accelerate product deterioration. Additional contamination and deterioration can be minimized by good manufacturing practices during repacking and maintaining the shellfish at low temperatures. If frozen shellfish are thawed during repacking, high bacterial counts and accelerated product deterioration may result. The requirement for maintaining shellfish meat temperatures at 7.2°C (45°F) or less during the entire repacking operation results in repacking having the most stringent sanitation requirements of all the certified processing and shipping operations.

Section G

Reshipping

Persons who obtain shellstock and shucked shellfish from certified dealers and sell the shellfish to other certified shippers or in interstate commerce shall be licensed and/or certified as shellstock shippers or reshippers. Use of the reshipper classification is left to the option of the state. The reshipper designation is most applicable to those dealers located at considerable distance inland or in inland states where obtaining shellfish from diggers or harvesters is unlikely.

Sanitation requirements for a reshipper depend upon the types of product handled and methods of operation. The recordkeeping requirements are particularly important. A reshipper shall not shuck or repack shellfish nor shall a reshipper remove or alter any existing label information. Appropriate additional information should be added to indicate the reshipper's name and certification number.

1. Reshipper Controls
Reshippers shall comply with all applicable requirements of Sections B-E.

Satisfactory Compliance
This item will be satisfied when:

a. All appropriate requirements of Sections B through E are met:

i. When shucked shellfish are handled, the following requirements apply:

. Section D.16, Source;

. Section D.22, Refrigeration of Shucked Shellfish;

. Section D.23, Ice; and

. Section D.24, Records.

ii. When shellstock is handled, the following requirements apply:

. Section B.1, Trucks;

. Section C.2, Wet Storage;

. Section D.1, Plant Location;

. Section D.2, Dry Storage and Protection of shellstock except that commingling shall not be permitted;

. Section D.5, Insect-control Measures;

. Section D.9, Plumbing and Related Facilities;

. Section D.10, Sewage Disposal;

. Section D.11, Poisonous or Toxic Materials;

. Section D.14, General Maintenance and Cleanliness;

. Section D.16, Source of Shellfish except that shellstock may not be obtained directly from harvesters;

. Section D.24, Records

iii. If a business consists only of a truck, a permanent business address where vehicles and records are available for inspection must be maintained and the following requirements apply:

. Section B.1, Boats and Trucks;

. Section D.16, Sources of Shellfish, except that shellstock may not be obtained directly from harvesters;

. Section D.22, Refrigeration and Shipping of Shucked Shellfish;

. Section D.24, Records;

b. The original labels on shucked shellfish and certified dealer's tags or labels on shellstock are maintained on the product containers. Labeling or tagging information is not altered or removed nor are shellstock commingled, resorted, or repackaged. The name and certification number of the reshipper shall be added to the package.

Public Health Explanation

This reshipper classification is intended to assist governmental agencies in maintaining control over the distribution of certified shellfish. The reshipper (RS) classification is an option which may be used by a state control agency in exercising adequate control.

Controls on reshippers are particularly important since shellfish dealers in inland states may wish to relabel, commingle, repackage, or otherwise make it difficult or impossible to determine the original source of the shellstock. These practices have interfered with many disease outbreak investigations and should be controlled by health authorities.

(This page is blank.)

Section H

Heat Shock

The heat shock method of preparing shellfish for shucking is used in several geographical regions and for several species of shellfish. The process is not intended to open the shellfish but rather to cause the shellfish to relax it's adductor muscle(s) so it can be more easily shucked. Short-term heat shock also resulted in a reduction of coliform and fecal streptococci numbers in the shucked meats and does not impair the keeping qualities of packed, refrigerated shellfish meats.

A variety of heat shock processes are currently in use and a large number of techniques are possible. Consequently, the Manual requirements are general in nature and emphasize the use of process schedules developed by or in cooperation with competent authorities. Other aspects of the process that require controls include washing of shellstock, cooling of heat shocked shellfish, refrigeration of heat shocked shucked shellfish, and cleaning of equipment. The requirements in this section pertain only to the heat shock process of the plant. In addition, the plant shall adhere to all other applicable requirements of this Manual.

1. Washing of Shellstock

Shellstock subjected to the heat shock process shall be washed with water from an approved source and culled of dead and damaged shellfish immediately prior to the heat shock operation. Shellstock shall be protected from contamination prior to and during the wash cycle.

Satisfactory Compliance
This item will be satisfied when:

a. All shellstock subjected to the heat shock process are washed with potable water of adequate supply and pressure; and culled of unsafe, unwholesome, and animals with badly broken shells immediately prior to the heat shock operation. Washing by immersion is prohibited.

b. Shellstock are handled in a manner which prevents their contamination during the pre-wash cycle.

Public Health Explanation

Although Section B.3 requires that shellstock be washed reasonably free of bottom sediments and detritus as soon after harvesting as is practicable, it is necessary to again wash and cull shellstock immediately prior to heat shocking to reduce the bacterial load in dipping tanks and on the surface of the shell. The cleaner the shellstock, the more rapidly the shellfish will arrive at the optimum temperature for shucking and there will be less variation in heat transfer among different lots.

One effect that heat shocking may have on shellfish is to leave the shells gaped due to a relaxed muscle. Consequently, it becomes difficult to distinguish dead shellstock which are gaped from live shellstock. Therefore, shellstock must be culled of dead animals prior to heat shock, otherwise they may be inadvertently shucked following heat shock.

2. Heat Shock Process

Scheduled processes for heat shock shall be established by the state shellfish control agency or by qualified persons, or in cooperation with qualified persons having adequate facilities for making such determinations and approved by the state shellfish control agency. The type, range, and combination of variations encountered in commercial heat shock operations shall be adequately provided for in establishing the scheduled process. Critical factors which may affect the scheduled process including species and size of the shellfish, time and temperature, and type of process shall be determined on the basis of a study of the effectiveness of the process. The scheduled process shall be such that the shellfish are not killed by the heat shock, the physical and organic properties of the shucked shellfish are not significantly changed, and the completed process does not cause additional microbiological deterioration of the shucked shellfish. Complete records covering all aspects of the scheduled process study and specifications shall be maintained by the person making the determination. Process schedules shall be posted in the plant.

Satisfactory Compliance
This item will be satisfied when:

a. An approved scheduled process is used in each heat shock processing plant and the scheduled process has been established by the SSCA or other qualified persons having adequate facilities for conducting appropriate studies to make such a determination.

b. Critical factors which may affect the process have been adequately studied and provided for in establishing the process. Critical factors to be considered include but are not limited to type and size of shellfish; time and temperature of exposure; type of process (e.g. hot water immersion, steam tunnel, steam retort); size of the tank, tunnel or retort; water-to-shellfish ratios in tanks and temperature and pressure recording devices.

c. The physical and organoleptic properties of the species are not changed by the scheduled process and the shellfish remain alive until shucked.

d. The process does not result in increased microbial deterioration of the shucked shellfish.

e. Records covering all aspects of the establishment of the process are maintained in the central file of the SSCA.

f. The scheduled process is posted at a conspicuous location in the plant and all responsible persons are familiar with the requirements.

Public Health Explanation

A variety of heat shock procedures including steam tunnels, retorts and hot water are acceptable if adequate studies have been conducted and measures are taken to ensure that the specific process will result in a safe product and additional microbial deterioration does not occur as a result of the process. The primary objective of heat shock is to facilitate shucking by subjecting the live shellfish to sufficient heat to cause the adductor muscle to relax and allowing the shucking knife to be easily inserted. An additional benefit is a reduction in the number of bacteria in the shucked meats (58, 59).

A 1963 study of heat shock process in South Carolina showed an overall reduction in the coliform and fecal coliform MPNs at all percentile levels. The greatest reduction occurred in the samples examined immediately after shocking. Holding on the shucking bench appears to result in a slight increase in these two groups of bacterial indices as compared to oysters examined immediately after shocking. However, these levels remain significantly lower than the levels obtained on samples from the cold shucking process (60). Another study of a steam tunnel process in Mississippi (59) confirms these results and concluded that with adequate time and temperature controls, a shucked product can be produced with lower initial bacterial levels and no detectable changes in flavor or texture.

Due consideration in developing the scheduled process must be given to a large number of factors which affects the heat shock process. Heat penetration into the shellfish will vary with species and size. Even regional variations in shell thickness and shape may affect the length of time required to reach the desired internal temperature. The temperature and time of exposure must be such that the adductor muscle is sufficiently relaxed to open easily but must allow the shellfish to remain alive. The scheduled process may be developed from studies conducted by the state, by a knowledgeable processor in cooperation with state shellfish control authorities, by shellfish experts such as university biologists or any other person with adequate knowledge of the technical control procedures. The person responsible for developing the scheduled process should retain all records of process operations so they may be reviewed by the FDA and state shellfish control authority if questions arise regarding the adequacy of the scheduled process or its use.

3. Cooling of Heat Shocked Shellstock

As part of the hot dip process, shellstock shall be subjected to an immediate cool down with water from an approved source.

Satisfactory Compliance
This item will be satisfied when:

a. All hot dipped shellstock are cooled immediately after the heat shock process. Dipping in an ice bath or use of flowing potable water shall be satisfactory for such cooling. The SSCA shall ensure that control procedures are in place to ensure that the ice dip process does not increase the potential for microbial contamination of the shellstock.

b. All heat shocked shellstock are handled in such a manner as to preclude contamination during the cooling process.

Public Health Explanation

After undergoing the heat shock process, the internal temperature of the meats are elevated to temperatures in excess of 37.8°C (100°F) (59). It is therefore necessary to reduce the internal temperatures of the shellfish meat immediately after the shucking process to prevent bacterial growth, but not to the extent that the purpose of the process is nullified.

4. Cooling of Heat Shocked Shucked Shellfish

The shucked meats from all shellstock which have been subjected to the heat shock process shall be cooled to an internal temperature of 7.2°C (45°F) within two hours after the heat shocking process. Refrigeration of containers of shucked shellfish in crushed ice is highly recommended.

Satisfactory Compliance
This item will be satisfied when:

All shellfish meats of shellstock which have been subjected to the heat shock process are shucked and cooled to at least 7.2°C (45°F) within two hours after the heat shock process and are placed in storage at 7.2°C (45°F) or below. This will require the use of crushed or flaked ice in the shucking containers, blowers, or chilling tanks, or the use of refrigerated water at the skimmer table.

Public Health Explanation

Shellfish meat temperatures of shellstock which have been subjected to the heat shock process are higher than those of conventionally shucked shellfish (58, 59, 60). Therefore, it is necessary that such meats be cooled quickly to 7.2°C (45°F) after the heat shock process to deter bacterial growth.

5. Cleaning of Heat Shock Process Equipment

Heat shock tanks or retorts, conveyors, tunnels, conveyances and all other equipment used in the heat shock process shall be cleaned in accordance with the requirements for cleaning of equipment established by Section D.15. Heat shock process tanks or retorts shall be of such construction that they may be easily cleaned.

Satisfactory Compliance
This item will be satisfied when:

a. Heat shock tanks or retorts, conveyors, tunnels, conveyances and all other equipment used in the heat shock process shall be thoroughly cleaned and sanitized in such a manner and at such a frequency as to minimize the danger of contamination of the shellfish in accordance with Section D.15. When used on a continuous basis, such equipment shall be cleaned and sanitized on a predetermined schedule using adequate methods for cleaning. Tanks, conveyances, and equipment are thoroughly cleaned at the end of each day's operation.

b. If a heat shock water tank is used, it is completely drained and flushed at three-hour intervals or less in such manner that all mud and detritus remaining in the dip tank from previous dippings are eliminated.

c. All heat shock process tanks or retorts, conveyors, tunnels, conveyances and other equipment are of such construction that they may be easily cleaned.

Public Health Explanation

If the water, mud, and detritus were allowed to remain in the heat shock tank or retort or on other process equipment under declining temperature conditions, it would constitute an excellent medium for growth of bacteria. Unclean equipment can also lead to contamination of food-contact surfaces throughout the plant.

The cleaning requirements for each heat shock process may vary. The plant should develop a cleaning schedule that is integrated with their scheduled process and which is approved by the state. The minimum requirements include changing the water in dip tanks at least every three (3) hours and cleaning all equipment at the end of each day's operation. Emptying the tank or retort and cleaning it at the close of the day's operation will help insure that the next day's dipping operation will start under optimum conditions of cleanliness. More frequent cleaning may be required depending upon how dirty the heat shock area and equipment become, and whether there is the potential for the shellfish and food-contact surfaces in the plant to become contaminated.

(This page is blank.)

Section I

Depuration

Depuration is intended to reduce the number of pathogenic organisms that may be present in shellfish harvested from moderately polluted (restricted) waters to such levels that the shellfish will be acceptable for human consumption without further processing. The process is not intended for shellfish from heavily polluted (prohibited) waters nor to reduce the levels of poisonous or deleterious substances which the shellfish may have accumulated from their environment. The acceptability of the depuration process is contingent upon the SSCA exercising very stringent supervision over all phases of the process.

1. Administrative Procedures
The depuration process shall be under the effective supervision of the SSCA. There shall be a state-approved control plan which details procedures for: controlling harvesting from restricted areas and transporting to the depuration plant*; approving of plant design and operation, including subsequent changes; certifying and inspecting plants in accordance with the requirements of this Manual; and prohibiting interstate shipments in the event that nonconformities are found which compromise the validity of the process. A Memorandum of Understanding (MOU) shall be developed between appropriate agencies when more than one state agency is involved in the control plan.

Satisfactory Compliance
This item will be satisfied when:

a. The responsible SSCA has adequate laws, resources, and equipment to survey restricted harvesting areas, to control harvesting of shellfish from restricted areas, and to control transportation to the depuration plant in order to prevent shellfish from being diverted into the marketplace prior to processing.

b. All plans for construction or remodeling of depuration plants are reviewed and approved by the SSCA prior to commencing construction.

c. A written depuration control plan is prepared by the SSCA which details control procedures required under this Section. The plan is updated as necessary.

d. Each person who harvests shellfish for delivery to a depuration plant has a separate permit issued by the responsible SSCA. The permit specifies the terms of the permit and states that any violation of the terms may result in revocation of the

---

*A depuration plant is defined as a facility of one or more depuration units. A depuration unit is a tank or series or tanks supplied by a single process water system.

permit. Individuals working under the supervision of a person holding a valid permit may not be required to have their own permits. See Part I, Section F, Paragraphs 1.a and 1.b for additional requirements.

e. Plant design, construction and operation of the process are evaluated in accordance with the requirements of this Manual and approved by the SSCA prior to initial certification and upon any subsequent changes in design or operation.

f. Depuration facilities are inspected and certified in accordance with the procedures established in Section A.2.

g. Certified depuration plants are inspected at appropriate frequencies to assure that each plant is operating in compliance with this Manual and that the inspection rating score is being maintained. Upon finding any major nonconformity, immediate corrections are made, or the plant's operating permit is suspended or the certification is canceled until corrections are made in accordance with Section A.2.

h. The SSCA conducts an analysis of the plant processing data and other pertinent records at least monthly to verify that the process and controls are effective and that the final product criteria for the plant as a whole or for specific growing areas are being met. More frequent evaluations are conducted if necessary.

i. Adequate records for each depuration plant are maintained by the state control agency in a central file indefinitely for items i, ii, and iii; and for a minimum of two (2) years for items iv, v, and vi below:

    i. the state approved depuration control plan;

    ii. the sanitary survey reports and data for the harvest areas;

    iii. the pre-certification process verification data and reports that are used to establish the operational specifications;

    iv. the periodic analysis of the process data;

    v. sanitary inspection reports for the facility; and

    vi. an evaluation report verifying that the operator's records have been reviewed and the process has been evaluated.

j. Shellfish are not transported interstate for the purpose of depuration unless an MOU is developed and implemented between the states to provide for adequate control measures to prevent diversion prior to processing.

k. When more than one (1) state agency is involved in the program, an MOU is developed which clearly defines control and administrative responsibilities.

Public Health Explanation

Extensive administrative procedures are essential if the state shellfish control agency is to adequately control a very complex operation such as depuration. There are numerous critical control points where significant deviation can result in the distribution of contaminated shellfish. Control over harvesting areas is needed to assure that the shellfish are not so contaminated that cleansing will be inadequate. Design, construction and operation of the plant must adhere to guidelines established in this section which are based upon studies conducted to verify that the process consistently produces safe shellfish. And finally, the inspection program must be adequate to detect critical deviations and to effect immediate correction or to prevent the sale of suspect shellfish.

2. Process Verification

A scheduled depuration process (SDP) shall be established by the state or by other qualified persons or in cooperation with qualified persons who have experience in designing and operating depuration plants. The process shall be based upon a comprehensive study of the effectiveness of the plant operations and shall be established prior to certification. The type, range, and combination of variations encountered in commercial operations shall be adequately addressed in the SDP. Critical control points shall be specified in the SDP. Records covering all aspects that form the basis for establishing the depuration process shall be retained in the central file of the SSCA.

Satisfactory Compliance
This item will be satisfied when:

a. A SDP is developed for each plant on the basis of experimental data which substantiates that the process adequately reduces the number of microorganisms present in shellfish harvested from restricted waters. The SDP shall undergo a plant evaluation study that will demonstrate the effectiveness of the process. The SDP will be reviewed and approved by the SSCA prior to the certification of each depuration plant.

The same SDP may be used by more than one (1) depuration plant, provided that an individual plant evaluation study is conducted, reviewed and approved by the SSCA for each plant.

b. The development of the SDP takes into account the following principles and procedures;

i. The SDP is developed by a person or in cooperation with persons knowledgeable in the design, construction, and operation of depuration plants.

ii. The scheduled process takes into account the critical process variables that may be encountered and be supported by studies showing that the process will consistently produce shellfish meeting end-point criteria. Specific process variables which shall be considered include, but are not limited to, shellfish species, seasonal effects, water temperature, salinity, dissolved oxygen (D.O.), turbidity, sources of the shellfish and process water, treatment of process water, tank design and construction, hydraulics, clearance between shellfish containers, clearance between shellfish containers and tank walls, processing time, raw product quality, end-point criteria, process monitoring, and general plant sanitation.

iii. For each depuration plant, the maximum allowable zero-hour level of fecal coliforms is determined during comprehensive process verification studies. Such studies shall be conducted over a sufficient period of time to encompass extremes of environmental conditions in the restricted harvest area and the depuration facility. The maximum zero-hour level of fecal coliforms in shellfish to be subjected to depuration shall be set such that the end-product criterion for depurated shellfish will be consistently met. Further evaluation of the zero-hour level for fecal coliforms shall continue during routine plant operations in order to insure that final (end-product) criteria are consistently met, despite changing environmental conditions in the restricted harvest area or the depuration plant. As a result of these continued studies, changes in zero-hour fecal coliform levels may be necessary.

c. Disposition of shellfish shall be as follows:

i. The SSCA can order the destruction or nonfood-use disposal of the shellfish used during the trial runs; or

ii The shellfish may be relayed in accordance with Part I, Section D; or

iii. Shellfish may be released to market only upon:

(a) Provisional certification by the SSCA; and

(b) Establishment of an appropriate sampling plan, based on the specific plant design and operational protocol, which has been formulated by the SSCA and concurred with by the U.S. Food and Drug Administration; and

(c) Samples are in compliance with Sections I.8.c and I.8.g.

d. Copies of all studies conducted in establishing the effectiveness of the SDP are retained indefinitely in the central file. The SDP or directions for carrying out the process are also maintained at the plant and are readily available to responsible individuals trained in the application of the process.

Public Health Explanation

Depuration is a complex biological process. Individual species respond in different ways to various combinations of operational criteria including water turbidity, salinity and temperature; depth of shellfish in the baskets; and tank design. There are numerous possible design and operation options (61, 62). Consequently, it is necessary to establish process efficiency for each plant, harvest area and species. The results of these studies are to be incorporated into a written SDP for each depuration plant and must specify the maximum level of fecal coliform bacteria in shellfish to be depurated, i.e., a maximum zero-hour criterion for fecal coliforms.

The SDP should be developed by a person or in cooperation with persons knowledgeable in the design, construction and operation of depuration plants. The scheduled process must take into account the critical process variables that may be encountered and be supported by studies showing that the process will consistently produce shellfish meeting end-point criteria. Specific variables which should be considered include, but are not limited to, shellfish species, water temperature, salinity, turbidity, sources of the shellfish and process water, treatment of process water, tank design and construction, hydraulics, processing time, raw product quality, end-point criteria, process monitoring, and general plant sanitation.

3. Plant Location, Design and Construction

The grounds about a depuration plant shall be free from conditions which may result in contamination of shellfish at any time during process and storage. The plant building or structure shall be suitable in size, construction, and design to prevent contamination of shellfish by animals and other pests; to keep untreated and treated shellfish separate; and to facilitate adequate cleaning, sanitizing, operation, and maintenance of the depuration facilities. Processing tanks, containers, piping, and conveyances are enclosed within a protective structure.

Satisfactory Compliance
This item will be satisfied when:

a. The grounds about a depuration plant meet the requirements of Section D.1.b.

b. Effective barriers, including roofs, screens, walls, and doors are provided where necessary to prevent the entry of animals and other pests into the process area.

c. Separate holding and storage areas are provided for purified and unpurified shellfish. Where needed to prevent cross-contamination, separate washing and culling facilities are provided for purified and unpurified shellfish and each facility has a conveniently located potable water supply.

d. Floors, walls, ceilings, lighting, heating and ventilation, water supply, insect control, plumbing, and sewage disposal comply, as necessary, with Section D.

Public Health Explanation

It is essential that depuration plants be designed and constructed so shellfish will be adequately protected and consistently purified. Research on the depuration process and experience gained in commercial facilities have led to some generally accepted standards which are critical for effective cleansing (31, 62, 63, 64). Other design and construction criteria are less clearly defined, and only general guidance is available. Additionally, the plant must be designed and constructed so adequate cleaning and sanitizing can be accomplished (36), and to facilitate proper operation.

4. Source of Shellfish

Shellfish intended for depuration shall be harvested only from growing areas meeting the water quality requirements for approved or restricted areas specified in Part I of this Manual. These areas shall produce shellfish which meet the zero-hour product specifications established in the process verification study. Shellfish destined for depuration plants shall be protected as necessary during harvesting and transporting to prevent contamination and undue physiological stress which could reduce the effectiveness of the depuration process. Adequate control measures shall be taken to prevent diversion of unpurified shellfish into the marketplace. Shellstock delivered to the depuration plant are properly identified with information necessary to trace each harvest lot back to the harvest area, date of harvest, and harvester or group of harvesters.

Satisfactory Compliance
This item will be satisfied when:

a. Shellfish intended for depuration are harvested only from waters which meet the NSSP requirements for restricted or approved growing areas. Shellfish harvested from approved or restricted areas meet the zero-hour product specifications established in the process verification study.

b. Shellfish subjected to depuration shall not be harvested from an area that yields shellfish exceeding the maximum zero-hour (pre-depuration) level of fecal coliform bacteria as determined for that plant under the SDP.

c. Strict and close supervision of harvesting and transportation of shellfish from restricted area is exercised to prevent diversion into the marketplace prior to depuration. This requires that harvest and transportation be under the continuing and effective supervision of the SSCA or their designated agent.

d. Shellfish are protected from contamination during harvesting and storage according to the requirements of Sections B and E of this Manual. Additionally, the

shellfish are protected from physical or thermal abuses which may reduce the effectiveness of the depuration process.

e. Shellstock intended for depuration are accompanied by complete and accurate harvest or transaction records specifying, at the minimum, harvest area (designated by the SSCA), harvester's permit numbers, date of harvest and amount shipped. Other information as required by the SSCA is included.

f. Each harvest lot of shellstock is designated, tagged, or marked in a manner acceptable to the SSCA to indicate that the shellfish are from restricted areas.

Public Health Explanation

It has been amply demonstrated that shellfish harvested from prohibited areas should not be used for depuration (31, 65, 66, 67). Depuration studies have been conducted on the relationship of initial levels of indicator bacteria and viruses to the levels of these indicators after varying lengths of time. These studies have indicated that consistent reductions of both bacteria and viruses to low levels can be achieved with moderately polluted shellfish, but satisfactory results cannot be obtained with heavily contaminated shellfish. Several of these studies (63, 66) further indicated that consistent cleansing was achieved when initial levels of fecal coliforms did not exceed 1,000 and 2,000 per 100 grams, respectively, for the eastern hard clam, *Mercenaria mercenaria* and the soft clam, *Mya arenaria*. Upper limits for oysters and other species of clams have not been established. Until such limits are established and adopted, field trials for each species and for various climatic conditions should be conducted as part of the process verification study. A maximum zero-hour criterion for fecal coliforms in shellfish to be depurated must be established.

Thermal and physical shock can adversely affect the pumping action of shellfish and reduce the rate of elimination of microorganisms (31, 32, 63). Contamination of the shellstock during harvest could raise the level of contamination to such a level that adequate cleansing will not occur. Thermal abuse also may cause bacterial levels to reach such levels that depuration may not be effective in 48 hours (68). The type of protection may include, but is not limited to, providing shade in warm weather, providing refrigeration in transit, insuring rapid transit to the depuration plant, preventing freezing in cold weather, preventing breaking of the shells, and optimizing holding or storage time before depuration.

It is also essential that shellfish harvested from restricted areas are controlled so they may not be illegally diverted and sold. This usually necessitates special procedures for monitoring harvest operations, and tagging the shellfish. Methods that may be employed include the use of specially designed, labeled, or colored containers; or the use of colored or distinctly shaped tags. If shellstock are transported in bulk, other methods to distinguish that the shellfish are from restricted areas may need to be employed. Recommended measures include continuous surveillance of the boat or truck, transporting in trucks sealed with a serially numbered tamper-evident seal, or a count by the SSCA of the quantity shipped and quantity received at the depuration plant.

5. Equipment Construction and Facility Design

Processing tanks, processing containers, and process water lines shall be designed and constructed from nontoxic materials and constructed so as to be easily cleanable. The processing system shall be designed and constructed so as to provide sufficient water of adequate quality throughout the system in a manner which accomplishes effective depuration.

Satisfactory Compliance
This item will be satisfied when:

a. Process tanks and plumbing lines are fabricated from non-toxic corrosion-resistant materials and are easily cleanable. Tanks and plumbing are constructed so that they are accessible for cleaning and inspection, are self-draining and are maintained in good repair. Plumbing is designed and installed so that physical or chemical cleaning and sanitizing will be effective.

b. Depuration tank design, dimensions, and construction is such that uniform hydraulic flow with minimum turbulence is maintained throughout each tank, and a means for easily measuring the flow rate of the process water* is provided for each tank. Clearance between shellfish containers and tank walls (especially the tank bottom) must be specified for each particular tank design and depuration water input system.

c. Shellfish containers have a mesh-type construction which allows water flow to all shellfish in the containers.

d. Process water meets physical, chemical, and microbiological parameters required for safety and normal physiological activity of the shellfish species. These parameters include:

i. minimum D.O. at any point in the tank capable of sustaining normal physiological activity by the shellfish. D.O. levels are to be specified as percent saturation for a given salinity and temperature of depuration process water. In no case shall the D.O. be less than 50% of saturation.

ii. no detectable coliform organisms as measured by the standard five-tube MPN test for drinking water (25) or a test of equivalent sensitivity in the tank influent.

---

* Process water is defined as the water in depuration tanks during the time that shellfish are being depurated.

iii. salinity to be ±20% of the median salinity regimes of the harvest area as determined by the SSCA. Salinities outside this range may be used if justified by data obtained during process verification studies.

iv. temperature to be a minimum of 10°C (50°F) for oysters and hard clams and 2°C (35°F) for soft clams. Temperature to be a maximum of 20°C (68°F) for hard and soft clams and 25°C (77°F) for oysters. Temperatures outside of these specified ranges can be used where shellfish are adapted to higher or lower temperatures. In such cases, maximum and minimum temperatures different from the values specified above must be determined during process verification studies. These studies must establish that the alternative temperatures promote normal physiological activity and efficient depuration.

v. pH to be 7.0–8.4.

e. A water treatment system is installed, where necessary, to provide an adequate quantity and quality of water for operating the depuration process. The treatment does not leave residues that may interfere with the process. The quality of the water prior to final disinfection shall meet the requirements of restricted or approved growing areas. Turbidity does not exceed a value which will inhibit normal physiological activity and/or would interfere with process water disinfection. The maximum allowable level of turbidity must be established for each plant during process verification studies.

Public Health Explanation

Processing tanks and containers used to hold shellfish which have cracked, rough or inaccessible surfaces, or which are made of improper material are apt to harbor accumulations of organic material in which bacteria, including pathogens, may reside and grow. Such organisms will regularly be introduced into the system from contaminated shellfish. Surfaces must be smooth and easily cleanable if bacteria are to be flushed out in the cleaning and sanitizing process. Uncleanable surfaces can result in inconsistent depuration effectiveness, and, possibly, the reintroduction of pathogens into the shellfish.

In addition, critical factors, such as water chemistry, tank dimensions, rate of flow, tank loading, and clearance between shellfish containers and tank walls must be controlled if the shellfish are to pump effectively and eliminate pathogenic organisms that may be present (62, 64, 69). The source of water and the water treatment system must be such that an adequate volume and quality of water can be provided to accomplish effective depuration. Currently, all plants in the United States use ultraviolet light for disinfection of process water (69, 70). Numerous studies have shown UV treatment to be highly effective for inactivating bacteria and viruses provided the units are properly maintained. In choosing a UV treatment system, consideration should be given to whether any pretreatment that may be needed to reduce turbidity, whether the process water will be recirculating or flow through, and whether the type of plant and flow rates are compatible with the UV treatment system (62, 64, 69, 70, 71, 72, 73).

Ozone has been used for many years in Europe for treating depuration process water. Care must be taken in using ozone or other chemicals which may react with organic and inorganic components of the water supply and form compounds which adversely affect physiological activity (69, 74). The use of ozone as a disinfectant for depuration process water is not currently sanctioned in the United States because there is a paucity of health effects data associated with this application. Therefore, disinfection with ozone and other chemicals may constitute a food additive situation requiring FDA approval before use. FDA has developed interim approval criteria for the use of ozone in both wet storage and depuration operations which include required assurances that free ozone is never in contact with shellfish (74).

6. Plant Operations

Operating procedures for each plant shall follow an established SDP which has been developed in accordance with Section I.2. of this Manual. Operating procedures shall also comply with applicable sanitation requirements of Sections B through F of this Manual and with the Current Good Manufacturing Practices in 21 CFR 110. Additionally, shellstock shall be washed and culled prior to processing, and handled and stored such that physiological conditions are not adversely affected and bacteriological quality does not deteriorate. Purified and unpurified shellfish shall be held in separate storage facilities. Processing water is provided in sufficient quantities to assure effective treatment of the shellfish. Processing time shall be adequate to achieve effective depuration.

Satisfactory Compliance
This item will be satisfied when:

a. Written operating procedures follow an established SDP which has been developed in accordance with Section I.2.; and includes the applicable requirements in Sections B through F of this Manual, and the Current Good Manufacturing Practices in 21 CFR 110.

b. Shellfish are thoroughly washed prior to processing with safe flowing water meeting the requirements of Section B.3.b. Shellfish are culled to remove dead, broken, or cracked shellfish, and the culled shellfish are destroyed or disposed of in such a manner as to prevent their use for human food. While awaiting depuration, shellfish are stored and handled in such a manner that physiological activity is not adversely affected and bacteriological quality does not deteriorate.

c. Shellfish from approved areas are not held or processed in the same plant in which shellfish from restricted areas are being depurated unless the approved area shellfish are also depurated.

d. Different harvest lots of shellfish are not commingled during washing, culling, processing, or packing. If more than one harvest lot of shellfish are being processed at the same time, the identity of each harvest lot is maintained throughout the depuration process and final packing. If shellfish in different tanks are at different stages of depuration, each tank is labeled to show when depuration began. Unpurified shellfish may not be removed from a depuration plant except under the direct supervision of the SSCA or approved agent.

e. Different shellfish species are not processed in the same depuration unit* unless studies demonstrate that the species are compatible.

f. A process verification study has established that tank loading rates, flow rates, dissolved oxygen and other parameters in the SDP will be adequate to insure sufficient physiological activity of the shellfish for depuration to occur at any point in the tank under maximum loading conditions. Container design and arrangement in the tank will result in sufficient uniform hydraulic flow around the shellfish.

g. The minimum flow rate of process water in each depuration tank shall be 107 liters per minute per cubic meter of shellfish (one gallon per minute per standard U.S. bushel of shellfish). The minimum volume of process water in depuration tanks shall be 6400 liters per cubic meter of shellfish (8 cubic feet of water per bushel) for hard clams and oysters and 4000 liters per cubic meter (5 cubic feet per bushel) for soft clams based upon the effective empty tank volume.** [Deviations from these criteria can be allowed only if process verification studies show that efficient depuration and end product bacteriological criteria are consistently obtained].

h. The maximum depth of shellfish in containers shall be 7.6 cm (3 inches) for hard clams and oysters, 20.3 cm (8 inches) for soft clams, and shall be determined by experimental data for other species. [The maximum depths specified may be increased if justified by data from process verification studies].

i. Shellfish are purified for at least 48 hours. Longer depuration times may be necessary under certain conditions.

Public Health Explanation

Since effective depuration is dependent upon the control of a wide range of interrelated variables, it is essential that a SDP be used in each plant. Operation of depuration plants in accordance with good manufacturing practices (36) is also necessary to assure that shellfish are effectively purified and are not recontaminated during subsequent processing and distribution.

---

* A depuration unit is a tank or series of tanks supplied by a single process water system.

** Effective empty tank volume is defined as the wetted volume of the unloaded tank to the overflow level.

Washing of shellstock prior to depuration rids shells of sand, mud, and detritus that may interfere with depuration and may make tank cleaning difficult. The type of harvest method may negate the need for additional washing. At other times, thorough washing at the plant may be necessary to adequately remove mud. After depuration, washing removes feces and pseudo-feces which may cling to shells and may recontaminate the shellfish meats during processing or consumption (62, 64, 69).

Simultaneous depuration of a shellfish species from restricted areas and processing or holding shellfish of the same species from approved areas in the same physical facility creates difficult control problems which increases the likelihood that contaminated shellfish may not be processed. In recognition of this problem, the ISSC at their 1984 meeting recommended against this practice (73). This does not preclude simultaneously processing or holding other species in the same facility under a separate certification number or processing or holding the same species from approved areas where there is an adequate time interval to assure that all restricted area and depurated shellfish have been removed from the premises, and there can be no commingling of purified and unpurified shellfish.

During design and construction of depuration systems, careful consideration must be given to hydraulic flow through the tank. Non-uniform flow may result in dead spots and oxygen depletion which lead to inadequate depuration at some locations in the tank. Choice of design criteria may be based on existing process verification studies or new studies which verify effectiveness of any new designs. Furfari (62) reports accepted design criteria for tank loading rates, water, flow and container arrangement. He recommends that tank water volume be at least 6400 liters per cubic meter of shellfish (8 cubic feet of water per U.S. bushel) for hard clams and eastern oysters and 4000 liters per cubic meter of shellfish (5 cubic feet per bushel) for soft clams. A minimum flow rate of 107 liters per minute per cubic meter of shellfish (1 gallon per bushel) is recommended to maintain adequate oxygen levels. He further recommends that there be at least 7.6 cm (3 inches) clearance separating containers of shellfish in the tanks and between the containers and the bottom and sides of the tank.

Different lots of shellfish must not be commingled in order to preserve lot identity and integrity. Handling and recordkeeping practices should be such that health officials can trace any container of shellfish that leaves the depuration plant back to a specific depuration cycle and to a specific harvest area. When shellfish do not adequately depurate, it is normally because the quality of the growing area has deteriorated and the bacteriological loading in the untreated shellfish is too high, or there was some breakdown in the process such as inadequate process water disinfection or inadequate flow rates.

The unpurified shellfish must be kept separate from purified shellfish to prevent accidental commingling of unprocessed shellfish with processed shellfish. Storage of unprocessed shellfish should be such that additional contamination does not occur and the physiological activity is not so adversely affected that depuration might not occur (62, 64, 69). Refrigeration may be required to provide this control.

Currently, the method of choice for disinfecting depuration process water in the United States is UV light treatment (69, 70). As with any disinfection system, microbial inactivation is strongly dependent on the dose-time relation which, for UV treatment, is primarily a function of water depth and turbidity. Contact time is a function of flow rate of water and cross-sectional area or volume of the unit. In order for the UV lights to remain effective, they must be kept clean to prevent build-up of materials which reduce radiation intensity. The amount of radiation must be monitored and the UV tubes replaced when they are no longer effective.

The results of previously conducted depuration studies indicate that minimum processing time is 48 hours under optimum conditions, but that under some conditions longer periods may be required for adequate depuration (64, 68, 69, 71, 72). Processing periods of less than 48 hours are not permitted. Conditioning shellfish from approved areas in tanks for less than 48 hours is covered under Section C on wet storage.

7. Maintenance and Cleaning

Only those chemical agents necessary for plant operations shall be present in the plant and shall be used only in accordance with labeling. Physical facilities of the plant including the processing system shall be kept in good repair, and are cleaned and sanitized as necessary. Cleaning and sanitizing agents shall be safe and effective for their intended use. No miscellaneous equipment is stored in processing or holding areas.

Satisfactory Compliance
This item will be satisfied when:

a. Use of chemical agents meets the requirements of Section D.11.

b. Cleaning of physical plant facilities including walls, ceilings and floors; and storage of equipment meets the requirements of Section D.14 and 15 b. and c.

c. Processing tanks and containers are cleaned, sanitized and rinsed free of the sanitizing agent before a batch is processed. When needed, processing tanks and containers are also cleaned after 24 hours of processing in a manner that will not recontaminate the shellfish.

d. As necessary to prevent contamination of the tanks and water, piping and reservoirs holding incoming water are cleaned, sanitized and rinsed free of sanitizing agent.

e. Disinfection units for the water supply are cleaned and serviced as frequently as is necessary to assure effective treatment of process water.

Public Health Explanation

The need to effectively clean and sanitize processing tanks and containers and pipes carrying process water is well established (63, 64, 73). The inadequate cleaning and sanitizing of process equipment can result in microorganisms being resuspended in the process water and increasing the bacterial loading to such a level that adequate depuration will not occur. The required frequency for cleaning and bulb replacement will vary depending upon such factors as water temperature, turbidity, and growth of fouling organisms.

8. Quality Assurance

An adequate routine sampling program shall be developed and maintained to assure: (1) that the plant continues to operate in accordance with the SDP; and (2) that each process batch of shellfish produced by the process meets the current end-point criteria established for the shellfish species being processed. Sample analyses shall be conducted by a laboratory approved by the SSCA pursuant to the requirements of Part I, Section B of this Manual. All quality assurance records shall be maintained for at least two (2) years to facilitate inspection by the responsible SSCA and review by FDA.

Satisfactory Compliance
This item will be satisfied when:

a. A routine sampling program is developed and applied to each process batch of shellfish to document that a harvest lot has been adequately cleansed and that the end product criteria are met. The sampling program provides for an adequate number of shellfish samples collected from representative locations within the tank and analyzed prior to and after depuration to assess the effectiveness of the process. The minimum sampling schedule shall be in accordance with the following table:

MINIMUM SAMPLING SCHEDULE

	Minimum Number of Samples		
No. of Harvest Areas	**Pollution Variability[2]**	**Incoming Shellfish**	**Final Product[1]**
Single Area	Low	Periodic[3] single samples	Single samples of each harvest lot[4] in each process batch.[5]
Single Area	High	One sample from each harvest lot	Duplicate samples of each harvest lot in each process batch.
Multiple Areas	Variable	Periodic single sample from each area	At least one duplicate sample from each highly variable area each week in each processing batch. Single samples of each process batch at other times during the same week.

[1] If final product criteria are met prior to 48 hours, the product may be released after 48 hours of depuration without a finished product sample.

[2] *Pollution variability* is determined by sanitary survey and water quality data evaluation as part of the SDP.

[3] The frequency of sampling is determined by the SSCA based upon the information from the SDP.

[4] A *harvest* lot is defined as all shellfish harvested from a particular area at a particular time and delivered to one depuration plant. The designation of areas is left to the SSCA.

[5] A *process batch* is a quantity of shellfish used to fill each separate depuration unit. A depuration unit is a tank or series of tanks supplied by a single process water system.

b. At the discretion of the SSCA, zero hour samples of unprocessed shellfish and samples of partially processed shellfish may be taken midway through the depuration cycle and analyzed for the end-point indicator. Such sample results may be used to predict whether the shellfish are acceptable for release. This approach can be used if an appropriate statistical analysis, performed or approved by the SSCA, of the samples collected during the process verification study and other historical sample results demonstrates this is feasible. Appropriate criteria and procedures are established to assure that the shellfish will be adequately cleansed when released. These beginning and mid-process samples must be supplemented with routine end-product samples to enable continuous verification of the reliability of the criteria.

c. The continuing evaluation of depuration plant performance and process efficiency is measured on the basis of end-point bacteriological assays of the depurated product using the geometric mean and upper 10% level (i.e., no more than 10% of the samples analyzed can exceed this value) for the species listed in the following table. In determining these values all of the final product data from every group of 10 consecutive process batches is used. For plants processing less than 10 batches in a three-month period, if any *one* sample exceeds the upper 10% level or if the geometric mean is in excess of the value specified, the plant is not in compliance.

End-Product Standards For Overall Depuration Plant Performance Evaluation

Species	Fecal coliforms per 100 grams	
	Geometric Mean[1]	Upper 10%
Soft Clam (*Mya arenaria*)	50	130
Hard Clam (*Mercenaria mercenaria*)	20	70
Oyster (*Crassostrea virginica*)	20	70

d. More intensive sampling of specific harvest lots is required by the SSCA when the shellfish are from a new harvest area and the effectiveness of depuration of those shellfish has not been established; and when the shellfish are from a harvest area which has historically exhibited highly variable water and/or shellfish quality, and end-product results are at times marginal or unsatisfactory.

[1] Each sample having an indeterminate low value will be assigned a value of 10 for computational purposes.

e. Untreated and treated process waters are sampled daily to test for critical parameters and to demonstrate that the water treatment units in use are operating effectively.

f. Samples are analyzed in a laboratory which has been evaluated and approved pursuant to the requirements of Part I, Section B. Laboratory analyses shall be performed by a state laboratory or a laboratory approved by the SSCA and shall use APHA, AOAC or ISSC approved methods to assure uniformity and acceptance of results. The laboratory shall be inspected as part of the annual plant certification and reinspected as necessary to assure that standard analytical methods are being applied. Any major changes in the laboratory personnel or facilities shall be followed by a reinspection. Laboratory inspections shall be conducted by state laboratory evaluation officers because of the special technical expertise required.

g. End-product standards for each process batch are indicated below. Shellfish from single process batches are not released to market unless laboratory results confirm that the end-point fecal coliform (FC) criteria established by the SSCA and included in the SDP are met at 24 or 48 hours for each process batch. The number of samples to be analyzed is based upon the information from the SDP. The following criteria have been established.

End Product Standards For Each Process Batch Of Shellfish
Fecal Coliforms per 100 grams

No. of Samples	Shellfish Species	Geo. Mean Not to Exceed	One Sample May Exceed	No Sample To Exceed
1	S.C.	-	-	170
	O., H.C.	-	-	100
2	S.C.	125	-	170
	O., H.C.	75	-	100
3	S.C.	110	-	170
	O., H.C.	45	-	100
5	S.C.	50	100	170
	O., H.C.	20	45	100
10	S.C.	50	130	170
	O., H.C.	20	70	100

S.C. = Soft Clam; O = Oyster; H.C. = Hard Clam; - denotes no value

h. All quality assurance records shall be maintained for at least the preceding two (2) years at both the depuration plants and at the laboratory.

Public Health Explanation

If shellfish are released for sale before the shellfish are adequately purified, adulterated shellfish may reach the market. It is therefore essential to develop an adequate sampling program designed to determine if critical environmental conditions are suitable for depuration and the shellfish being released to the market meet accepted criteria. A minimum sampling schedule is specified. However, due to variations in plant size and design, it may be necessary to increase sampling. Consequently, a sampling program must be developed for each plant. Based on past experience, the sampling program should include at least the following (1) total coliform level of the raw water source and treated water in the tank, (2) fecal coliform levels of incoming (zero hour) and final product (end-product) shellfish as specified above, (3) temperature, turbidity, and salinity of the process water, and (4) dissolved oxygen levels at the tank outlet.

Extensive field work with soft clams has established a NSSP recommended end-point for that species (8). For the hard clam and eastern oyster, studies have indicated that a lower end-point criteria are desirable and that consistent depuration to non-detectable levels of fecal coliforms is feasible. At the 1985 meeting of the ISSC interim end-point criteria for hard clams and eastern oysters were adopted and the continuing use of the end-point criteria for soft clams was endorsed (75). Release may also be based upon mid-point sample results provided the depuration system has established a proven record. For other species, it is recommended that end-point criteria be developed on a case-by-case basis for each scheduled process, and that it be predicated on the level of indicators found in approved growing areas, in studies conducted while establishing the SDP process and in existing scientific literature on depuration.

9. Packing and Labeling

Depurated shellfish shall be washed and culled after depuration and packaged in clean containers fabricated from safe materials. Different harvest lots of shellfish shall not be commingled during packing. Each container of depurated shellstock shall have securely affixed to it a durable, waterproof tag or label bearing information that will identify the shellfish as being depurated and will allow the shellfish to be traced back to a specific batch and specific harvest lot.

Satisfactory Compliance
This item will be satisfied when:

a. Shellfish are washed and culled after depuration, and packaged in clean shipping containers fabricated from safe materials. The depurated shellfish shall not be repackaged after leaving the depuration plant in order to maintain identity. Use of tamper-evident seals is recommended.

b. Shellfish from different harvest lots are packaged separately.

c. The shellfish are protected from contamination and are held at an ambient temperature no higher than 10°C (50°F). Shellfish are not permitted to remain unrefrigerated for prolonged periods during warm weather.

d. Durable, waterproof tags of minimal size - 6.7 by 13.3 cm (2 5/8 by 5 1/4 inches) - are securely affixed to each package of depurated shellstock. These tags are approved by the SSCA and include at least the following information in legible and indelible form:

   i. name, address and DP certification number of the processor;

   ii. date of processing;

   iii. depuration cycle number;

   iv. the identity of the harvest area or aquaculture site;

   v. type and quantity of shellfish; and

   vi. the following statement will appear in bold capitalized type "THIS TAG IS REQUIRED TO BE ATTACHED UNTIL CONTAINER IS EMPTY AND THEREAFTER KEPT ON FILE FOR 90 DAYS."

Public Health Explanation

Depurated shellfish require an increased level of control compared to shellfish from approved areas because of the increased potential for contamination. These controls must include packaging and labeling that will serve to help identify the depuration cycle of each harvest lot and to deter illegal commingling of undepurated shellfish with depurated shellfish. Such controls include prohibition against commingling of harvest lots during packing, tags that identify the shellfish as being depurated, and a prohibition against repackaging after the shellfish leave the depuration plant. It is recommended that tamper-evident seals be used on the packages as a further deterrent.

10. Supervision and Recordkeeping

Management shall clearly designate a knowledgeable and competent individual to be present at the plant and be accountable that operating procedures and proper personal hygiene practices are followed. The supervisor will also maintain complete and accurate records that will permit each package of depurated shellfish to be traced back to its source, and will account for all product sample results and measurements of critical parameters for each cycle.

Satisfactory Compliance
This item will be satisfied when:

a. A reliable, knowledgeable and competent individual has been designated by management to supervise the depuration operations to insure that all operating procedures are followed, and that all shellfish leaving the depuration plant meet product specifications. Supervisory personnel have a thorough knowledge of the process and are present during critical process operations.

b. There is evidence that the supervisor has been monitoring employee hygiene practices, and insuring that proper plant sanitation practices are implemented, including plant and equipment clean-up and maintenance.

c. A record is established for each depuration cycle or run. To accomplish this, each cycle is assigned a sequential number so that monitoring and sampling data for that cycle can be correlated with the cycle number. A cycle may include more than one harvest lot of shellfish provided all lots begin the depuration cycle at the same time and the conditions of depuration are the same, and harvest lot separation is maintained.

d. Complete and accurate records must be maintained by the depuration processor. The necessary information and the required standard format for maintaining records of purchases and sales is illustrated in Appendix D. These records are sufficient to document a package of depurated shellstock and trace it back to its process batch, harvest lot or area, date of harvest and harvester. Purchases are recorded in a permanently bound ledger book and are maintained for a minimum of one (1) year. Results of product samples and measurement of critical parameters are maintained for a minimum of two (2) years. If records are maintained on a computer, they must contain all the necessary information contained in Appendix D.

e. A current copy of the plant operating procedures including critical parameters and end-product specifications is maintained at the depuration plant.

Public Health Explanation

It is essential that detailed identification information be maintained on all harvest lots and shipping containers of depurated shellfish. In the event a disease outbreak occurs or a question arises concerning the product, responsible state and federal agencies must be able to trace the implicated shellfish back to a specific depuration cycle, and to the harvest area. Additionally, maintaining complete and accurate records of all transactions serves to promote business integrity wherein all harvesters, processors, and dealers are fully accountable for their product. Records of product samples and critical parameters within the plant are necessary to determine if the plant is operating in accordance with the SDP. Their records should be kept for at least two (2) years in order that adequate investigations can be conducted in the event of a suspected illness and in order that process reviews can be made by the state shellfish control authority.

SHELLFISH INDUSTRY

Equipment Construction Guides

Developed for use with the
Public Health Service-States-Industry
Cooperative Program for the
Certification of Interstate Shellfish Shippers

Compiled and Edited by
Lee Roy Lunsford

U. S. DEPARTMENT OF HEALTH, EDUCATION, AND WELFARE
Public Health Service
Division of Environmental Engineering and Food Protection
Washington 25, D. C.

INTRODUCTION

Since 1925 the Public Health Service, the States, and the shellfish industry have cooperated in a program designed to maintain a high level of sanitation in the growing, harvesting, and processing of oysters, clams, and mussels to be marketed as a fresh or frozen product. The basic sanitary standards used in this program are fully described in PHS Publication No. 33, Manual of Recommended Practice for Sanitary Control of the Shellfish Industry, Parts I and II. General construction and cleanability standards for equipment used by the shellfish industry are an integral part of these basic standards.

The need for more specific construction guides for equipment used by the shellfish industry was reviewed at the 1958 Shellfish Sanitation Workshop,[1] and the Public Health Service was requested to initiate development of such guides. In accord with the request, the Public Health Service developed initial drafts of equipment construction guides. Agencies and organizations which received these initial drafts and thus contributed to the development of the completed construction guides included: Interested State agencies, Oyster Institute of North America, Bureau of Commercial Fisheries, Food and Drug Administration, Canadian Department of National Health and Welfare, and two equipment manufacturing companies. The completed construction guides were reviewed and adopted by the 1961 National Shellfish Sanitation Workshop.

It is the purpose of this guide to describe construction and fabrication procedures which will insure that blower tanks, skimmers, returnable shipping containers, and shellfish shucking buckets and pans will meet the equipment construction standard of the Cooperative Program and also the functional needs of industry. However, the development of new methods of equipment construction or fabrication is also encouraged. Therefore, shellfish equipment specifications heretofore or hereafter developed which so differ in design, material, fabrication, or otherwise as not to conform with the following standards, but which in the fabricator's opinion are equivalent to or better may be submitted for consideration. Correspondence relative to such standards should be directed to Shellfish Sanitation Program, Division of Environmental Engineering and Food Protection, Public Health Service, Washington 25, D.C.

SCOPE

This guide covers the sanitary construction aspects of (1) shellfish blower tanks, including the sanitary piping for air, water, and drain lines, (2) the stand-supported skimmer, including the supporting stand, (3) returnable shipping containers, (4) shellfish shucking buckets, and (5) shellfish shucking pans.

DEFINITIONS

(1) *Shellfish*—All edible species of oysters, clams, or mussels. Shellfish products which contain any material other than the meats and/or shell liquor of oysters, clams, or mussels will be regarded as a "processed food" and will not be included in the Cooperative Program. (For the purpose of this guide, the term does not include crabs, shrimps, or lobsters.)

(2) *Shucked Shellfish*—Shellfish, or parts thereof, which have been removed from their shells.

(3) *Product*—Shucked shellfish which are either held, shipped, washed, and/or drained in this equipment.

(4) *Product Contact Surface*—All surfaces that are exposed to the product, or surfaces from which liquid may drain, drop, or be drawn into the product.

(5) *Non-Product Contact Surfaces*—All other surfaces not included in (4) above.

---

[1] Proceedings, 1958 Shellfish Sanitation Workshop, U.S. Public Health Service, Washington 25, D.C.

April 1962
640622—62

1

(6) *Blower*—A tank-like device for the immersion washing of shucked shellfish. Air may be introduced at the bottom of the tank to produce agitation.

(7) *Drain gate and chute*—The tank opening through which the washed shellfish are discharged.

(8) *Drain valve*—The valve through which the wash water is released to the floor or waste line.

(9) *Skimmers*—The stand-supported, perforated tray in which shucked shellfish are spray washed and/or drained.

(10) *Skimmer Paddle*—The utensil used as the gate on the skimmer exit chute and/or one used to scrape the product through the exit chute.

(11) *Returnable Shipping Container*—Multiple use containers for holding or shipping of shucked shellfish.

(12) *Shellfish Shucking Bucket*—Containers for temporarily holding shucked shellfish during the shucking process.

(13) *Shellfish Shucking Pan*—Containers for temporarily holding shucked shellfish during the shucking process.

(14) *Welds*—Permanent seams and joints.

BLOWER TANK

A. Material

(1) All product-contact surfaces shall be of A.I.S.I.[1] Type No. 302 stainless steel or equally corrosion-resistant metal that is nontoxic and nonabsorbent except that:

(a) Plastic materials may be used for the blower tank drain gate and drain valve. These materials shall be relatively inert, resistant to scratching, scoring, and distortion by the temperature, chemicals, and methods to which they are normally subjected in operation, or by cleaning and bactericidal treatment. They shall be nontoxic, fat resistant, relatively nonabsorbent, relatively insoluble, and shall not release component chemicals or impart a flavor to the product.[2]

(b) Rubber and rubber-like materials may be used for blower tank paddles or gate, drain gate, and drain valve. These materials shall be relatively inert, resistant to scratching, scoring, and distortion by the temperature, chemicals, and methods to which they are normally subjected in operation, or by cleaning and batericidal treatment. They shall be nontoxic, fat resistant, relatively nonabsorbent, relatively insoluble and shall not release component chemicals, nor impart a flavor to the product.[2]

(2) All nonproduct contact surfaces shall be of inherently corrosion-resistant material, shall be rendered corrosion-resistant, or shall be painted. Surfaces to be painted shall be effectively prepared for painting; and the paint used shall adhere, be relatively nonabsorbent, and shall provide a smooth, cleanable, and durable surface. Parts having both product-contact and non-product-contact shall not be painted.

B. Fabrication

(1) All product-contact surfaces shall be at least as smooth as No. 4 mill finish on stainless steel sheets.

(2) All seams in product-contact surfaces shall be welded with the welds ground smooth and polished to not less than a No. 4 finish. All outside seams shall be smooth and waterproof. All weld areas and deposited weld material shall be substantially as corrosion-resistant as the parent metal.

(3) All appurtenances, including drain gates and chutes having product-contact surfaces, shall be easily removable for cleaning, or shall be readily cleanable in place.

(4) All product-contact surfaces shall be easily accessible, visible, and readily cleanable, either when in an assembled position or when removed.

(5) All internal angles of 135° or less on product-contact surfaces shall have minimum radii of 1/4 inch, except that minimum radii for fillets or welds in product-contact surfaces may be smaller for essential functional reasons. In no case shall radii be less than 1/8 inch.

[1] American Iron and Steel Institute.

[2] Plastic, rubber, and rubber-like materials used for equipment may be subject to the Food Additive Amendment to the Federal Food, Drug, and Cosmetic Act. The acceptability of such materials under the Food Additive Amendment shall be obtained from equipment manufacturers.

2

April 1962

(6) All sanitary pipe fittings shall conform to "3–A Sanitary Standards for Fittings Used on Milk and Milk Products Equipment," and supplements thereto.[4]

(7) Nonproduct-contact surfaces shall have a smooth finish, be free of pockets and crevices, and readily cleanable.

(8) Legs shall be of sufficient length to provide at least 12 inches clearance between the lowest fixed point of the tank and the floor, shall be smooth with rounded ends, and shall have no exposed threads. If legs are of hollow tube stock, they shall be effectively sealed.

(9) All threads on product-contact surfaces shall comply with specifications for threads contained in the 3–A Sanitary Standards for Fittings.[4]

(10) External and internal sections of the air pipe shall be easily cleanable to a point at least two inches above the tank overflow level.

(11) The false bottom shall be so constructed as to be rigid and, in any event, of at least 16 U.S. standard gage stainless steel, or equivalent material.

(12) Perforations or slots in the false bottom shall be not less than $\frac{3}{16}$ inch in the minimum diameter and the end radius of the perforations shall be not less than $\frac{3}{32}$ inch. After perforation, the flat surface of the sheet from which the perforating punch or drill emerges on the down stroke shall be polished to the equivalent of not less than a No. 4 mill finish.

(13) Air lines shall be of easily cleanable construction to a point two inches above the tank overflow.

(14) Wire mesh shall not be used.

(15) The blower tank shall be constructed so that it will not buckle or sag and so that it will be self-draining. Product-contact surfaces shall be constructed of not less than 16 U.S. standard gage stainless steel or equivalent material.

(16) Maximum dimension of the tank from point of overflow to drain valve flange shall not exceed 40 inches.

(17) Drain valves and flange shall comply with the 3–A Sanitary Standards for Fittings used on Milk and Milk Products Equipment. The flange shall be welded to the body of the blower tank.

(18) There shall be no exposed screw, bolt or rivet heads on product-contact surfaces.

SKIMMERS

A. Material

(1) All product-contact surfaces shall be of A.I.S.I. type No. 302 stainless steel, or equally corrosion-resistant metal that is non-toxic and nonabsorbent, except that:

(a) Suitable plastic materials or rubber and rubber-like materials may be used for the skimmer paddle or gate. These materials shall be relatively inert, resistant to scratching, scoring, and distortion by temperature, chemicals, and methods to which they are normally subjected in operation, or by cleaning and bactericidal treatment. They shall be nontoxic, fat resistant, relatively nonabsorbent, relatively insoluble, and shall not release component chemicals nor impart a flavor to the product.[3]

(2) All nonproduct-contact surfaces shall be of inherently corrosion-resistant material, shall be rendered corrosion-resistant, and except for funnel drain, shall be painted. Surfaces to be painted shall be effectively prepared for painting and the paint used shall adhere, be relatively nonabsorbent, and shall provide a smooth, cleanable, and durable surface. Parts having both product-contact and nonproduct-contact surfaces shall not be painted.

B. Fabrication

(1) All product-contact surfaces shall be at least as smooth as a No. 4 mill finish on stainless steel sheets.

(2) All seams in product-contact surfaces shall be welded with the welds ground smooth and polished to not less than a No. 4 mill finish. All outside seams shall be smooth and waterproof. All weld areas and deposited weld metal shall be substantially as corrosion resistant as the parent metal.

[3] Plastic, rubber, and rubber-like materials used for equipment may be subject to the Food Additive Amendment to the Federal Food, Drug, and Cosmetic Act. The acceptability of such materials under the Food Additive Amendment shall be obtained from equipment manufacturers.

[4] Sanitary standards describing the construction of valves, fittings, and pumps may be obtained from International Association of Milk and Food Sanitarians, Inc., Box 347, Shelbyville, Indiana.

April 1962

3

(3) All appurtenances having product-contact surfaces shall be easily removable for cleaning, or shall be readily cleanable in place.

(4) All product-contact surfaces shall be easily accessible, visible, and readily cleanable, either when in an assembled position or when removed. The skimmer shall be demountable from the supporting stand for cleaning.

(5) All internal angles of 135° or less on product-contact surfaces shall have minimum radii of 1/4 inch, except that minimum radii for fillets of welds in product-contact surfaces may be smaller for essential functional reasons.

(6) The skimmer shall be constructed so that it will not buckle or sag while in use, so that both the perforated area and drainage funnel are self-draining, and so as to provide plane surfaces free from depressions, indentations, or bulges which prevent draining when the pitch is not greater than 1" in 50". (Corners and rims of perforated skimmers should be adequately reinforced to prevent damage from handling during cleaning and bactericidal treatment.)

(7) The product-contact surfaces shall be constructed of not less than 16 U.S. standard gage stainless steel or equivalent material. The perforations or slots in the strainer shall be at least 1/4" in diameter (dimension A, Fig. 1) and not more than 1¼" apart (dimension B, Fig. 1).[1] The strainer area shall have no perforations within 1/2" of the edge (dimension C, Fig. 1).[2] After perforations, the flat surface of the sheet from which the perforating punch or drill emerges on the down stroke shall be polished to the equivalent of not less than a No. 4 mill finish. No bracing for the skimmer or the skimmer support stand shall block any perforations unless the brace is made of corrosion-resistant material and fabricated in a manner suitable

for a product-contact surface, and unless it can be readily removed for cleaning. A minimum of 3½" shall be provided between the strainer and the top of the skimmer (dimension E, Fig. 1).

(8) Legs shall be smooth with rounded ends, and have no exposed threads. If legs are of hollow tube stock, they shall be effectively sealed.

(9) A minimum vertical clearance of 2" shall be provided between the perforated skimmer area and the drainage funnel. (Dimension D, Fig. 1).

(10) There shall be no threads on product-contact surfaces except as provided for in the 3-A Sanitary Standards for Fittings.

(11) The funnel drain shall have a discharge opening of a size sufficient to discharge the drainage without pooling above, and be not less than equivalent to a diameter of 4". The funnel drain shall terminate in a free discharge, a distance of at least 6" above the floor or the drain connection if located at a higher elevation than the floor.

(12) Frames, frame legs, and supporting edge for the skimmer shall have:

    (a) Structural parts not in contact with the product, and parts constructed with a smooth finish so as to be readily cleanable.

    (b) Self-draining exterior surfaces.

    (c) A minimum of 6" of space between the lowest part of the frame and the floor to provide ready access for cleaning legs and feet and those parts not readily removable.

(13) The frame shall provide continuous support for the outside edge of the skimmer strainer.

(14) The receiving-container shelf under the skimmer chute, where provided as an integral part of the skimmer support frame, shall be constructed of nonabsorbent, corrosion-resistant material and located so that the receiving-container rim will be at least two feet above the floor.

(15) All seams in the funnel drain area shall be smooth and waterproof, and substantially as corrosion resistant as the parent metal.

(16) There shall be no exposed bolts, screws, or rivets in the product-contact surfaces.

April 1962

---

[1] Sanitary standards describing the construction of valves, fittings, and pumps may be obtained from International Association of Milk and Food Sanitarians, Inc., Box 347, Shelbyville, Indiana.

[2] Skimmer size: The Food and Drug Administration definition and standard of identity for raw oysters states in part: "The oysters are drained on a strainer or skimmer which has an area of at least 300 square inches per gallon of oysters drained, and has perforations of at least 1/4 of an inch in diameter and not more than 1¼ inches apart, or perforations of equivalent areas and distribution. (Definitions and Standards under the Federal Food, Drug, and Cosmetic Act, Title 21, Part 36, Federal Register, August 27, 1945)."

4

RETURNABLE SHIPPING CONTAINERS

A. Material

(1) All metallic product-contact surfaces shall be of A.I.S.I. type No. 302 stainless steel or Aluminum Association type No. 5052-0 alloy, or equally corrosion-resistant metal that is nontoxic. If constructed of stainless steel, the containers shall not be constructed of less than 20 gauge material. If constructed of aluminum alloy the material shall not have a thickness less than 0.064 inches.

(2) All nonproduct-contact surfaces shall be of corrosion-resistant material, and shall provide a smooth, cleanable, and durable surface.

B. Fabrication

(1) All product-contact surfaces shall be at least as smooth as a number 4 mill finish on stainless steel, or equivalent surface finish on aluminum.

(2) All internal angles of 135° or less on product-contact surfaces shall have minimum radii of ¼ inch.

(3) There shall be no seams, crevices, or other openings within the food-contact surfaces.

(4) The container rim shall be rolled so as to permit easy and complete cleaning. The bead shall either be an open type with an external radii of not less than 3/16 inch or a sealed closed type.

(5) The container lid shall be so constructed as to afford easy and complete cleaning, shall be reasonably tight fitting, and a lip shall extend at least one inch down the outside of the container. Provisions shall be made for sealing the container so that any tampering will be evident.

(6) Handles shall be provided on 5-gallon or larger containers. The handles shall be considered as a nonproduct-contact surface.

SHUCKING BUCKETS AND PANS

A. Material

(1) All metallic product-contact surfaces shall be of A.I.S.I. type No. 302 stainless steel or Aluminum Association type No. 5052-O aluminum alloy, or equally corrosion-resistant metal that is nontoxic. If constructed of stainless steel, the buckets shall not be constructed of less than 22 gage material and the pans shall not be constructed with less than 24 gage material, or if constructed of aluminum alloy, the material shall not have a thickness less than 0.064 inches.

(2) All nonproduct-contact surfaces shall be of corrosion-resistant material, and shall provide a smooth, cleanable, and durable surface.

B. Fabrication

(1) All product-contact surfaces shall be at least as smooth as a number 4 mill finish on stainless steel or equivalent surface finish on aluminum.

(2) All internal angles of 135° or less on product-contact surfaces shall have minimum radii of ¼ inch.

(3) The shellfish shucking bucket shall not exceed a nine-pint capacity, except for the soft clam (Mya arenaria) shucking pan which shall not exceed a four-pint capacity.

(4) There shall be no seams, crevices, or other openings within the food-contact surfaces, except that two holes 180° apart shall be permitted in the side of each bucket near the top to accommodate a removable ball-type handle.

(5) The container rim shall be so constructed as to afford maximum strength and protection against damage, and shall be so rolled as to permit easy and complete cleaning. The bead shall be open type with an external radii of not less than 3/16", or a sealed closed type.

(6) The bail, if provided, shall be considered as contact surface and subject to material specifications as outlined in paragraph A of this standard. The bail shall be not less than 3/16" in diameter; it shall be so constructed that it will be held into place by spring tension. The bail shall be so constructed that it can be easily removed from the shucking bucket for cleaning purposes.

APPENDIX A

SKIMMER DESIGN DETAIL

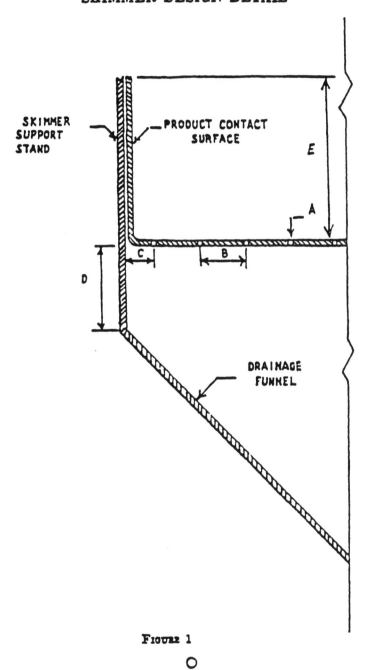

FIGURE 1

O

6   April 1962

Adopted by the 1961 National
Shellfish Sanitation Workshop

PUBLIC HEALTH SERVICE PUBLICATION NO. 943
First Printing
April 1962

Appendix B

Selected References in the Code of Federal Regulations
Applicable To Shellfish Processing and Products

PART 101—FOOD LABELING

Subpart A—General Provisions

Sec.
101.1 Principal display panel of package form food.
101.2 Information panel of package form food.
101.3 Identity labeling of food in packaged form.
101.4 Food; designation of ingredients.
101.5 Food; name and place of business of manufacturer, packer, or distributor.
101.6 Label designation of ingredients for standardized foods.
101.8 Labeling of food with number of servings.
101.9 Nutrition labeling of food.
101.10 Nutrition labeling of restaurant foods.
101.11 Saccharin and its salts; retail establishment notice.
101.13 Sodium labeling.
101.15 Food; prominence of required statements
101.17 Food labeling warning and notice statements.
101.18 Misbranding of food.

Subpart B—Specific Food Labeling Requirements

101.22 Foods; labeling of spices, flavorings, colorings and chemical preservatives.
101.25 Labeling of foods in relation to fat and fatty acid and cholesterol content.
101.29 Labeling of kosher and kosher-style foods.
101.33 Label declaration of D-erythroascorbic acid when it is an ingredient of a fabricated food.
101.35 Notice to manufacturers and users of monosodium glutamate and other hydrolyzed vegetable protein products.

Subparts C-E—[Reserved]

Subpart F—Exemptions From Food Labeling Requirements

101.100 Food; exemptions from labeling.
101.103 Petitions requesting exemptions from or special requirements for label declaration of ingredients.
101.105 Declaration of net quantity of contents when exempt.
101.108 Temporary exemptions for purposes of conducting authorized food labeling experiments.

Sec.
APPENDIX A TO PART 101—MONIER-WILLIAMS PROCEDURE (WITH MODIFICATIONS) FOR SULFITES IN FOOD. CENTER FOR FOOD SAFETY AND APPLIED NUTRITION, FOOD AND DRUG ADMINISTRATION (NOVEMBER 1985)

AUTHORITY: Secs. 4, 5, 6 of the Fair Packaging and Labeling Act (15 U.S.C. 1453, 1454, 1455); secs. 201, 301, 402, 403, 409, 701 of the Federal Food, Drug, and Cosmetic Act (21 U.S.C. 321, 331, 342, 343, 348, 371).

SOURCE: 42 FR 14308, Mar. 15, 1977, unless otherwise noted.

Subpart A—General Provisions

§ 101.1 Principal display panel of package form food.

The term "principal display panel" as it applies to food in package form and as used in this part, means the part of a label that is most likely to be displayed, presented, shown, or examined under customary conditions of display for retail sale. The principal display panel shall be large enough to accommodate all the mandatory label information required to be placed thereon by this part with clarity and conspicuousness and without obscuring design, vignettes, or crowding. Where packages bear alternate principal display panels, information required to be placed on the principal display panel shall be duplicated on each principal display panel. For the purpose of obtaining uniform type size in declaring the quantity of contents for all packages of substantially the same size, the term "area of the principal display panel" means the area of the side or surface that bears the principal display panel, which area shall be:

(a) In the case of a rectangular package where one entire side properly can be considered to be the principal display panel side, the product of the height times the width of that side;

(b) In the case of a cylindrical or nearly cylindrical container, 40 percent of the product of the height of the container times the circumference;

(c) In the case of any otherwise shaped container, 40 percent of the total surface of the container: *Provided, however,* That where such container presents an obvious "principal display panel" such as the top of a triangular or circular package of cheese, the area shall consist of the entire top surface. In determining the area of the principal display panel, exclude tops, bottoms, flanges at tops and bottoms of cans, and shoulders and necks of bottles or jars. In the case of cylindrical or nearly cylindrical containers, information required by this part to appear on the principal display panel shall appear within that 40 percent of the circumference which is most likely to be displayed, presented, shown, or examined under customary conditions of display for retail sale.

§ 101.2 Information panel of package form food.

(a) The term "information panel" as it applies to packaged food means that part of the label immediately contiguous and to the right of the principal display panel as observed by an individual facing the principal display panel with the following exceptions:

(1) If the part of the label immediately contiguous and to the right of the principal display panel is too small to accommodate the necessary information or is otherwise unusable label space, e.g., folded flaps or can ends, the panel immediately contiguous and to the right of this part of the label may be used.

(2) If the package has one or more alternate principal display panels, the information panel is immediately contiguous and to the right of any principal display panel.

(3) If the top of the container is the principal display panel and the package has no alternate principal display panel, the information panel is any panel adjacent to the principal display panel.

(b) All information required to appear on the label of any package of food pursuant to §§ 101.4, 101.5, 101.8, 101.9, 101.17, 101.25 and Part 105 of this chapter shall appear either on the principal display panel or on the information panel, unless otherwise specified by regulations in this chapter.

(c) All information appearing on the principal display panel or the information panel pursuant to this section shall appear prominently and conspicuously, but in no case may the let-

ters and/or numbers be less than one-sixteenth inch in height unless an exemption pursuant to paragraph (f) of this section is established. The requirements for conspicuousness and legibility shall include the specifications of §§ 101.105(h) (1) and (2) and 101.15.

(1) Packaged foods are exempt from the type size requirements of this paragraph: *Provided,* That:

(i) The package is designed such that it has a surface area that can bear an information panel and/or an alternate principal display panel.

(ii) The area of surface available for labeling on the principal display panel of the package as this term is defined in § 101.1 is less than 10 square inches.

(iii) The label information includes:

(a) Nutrition labeling in accordance with § 101.9.

(b) A full list of ingredients in accordance with regulations in this part and the policy expressed in § 101.6.

(iv) The information required by paragraph (b) of this section appears on the principal display panel or information panel label in accordance with the provisions of this paragraph (c) except that the type size is not less than three sixty-fourths inch in height.

(2) Packaged foods are exempt from the type size requirements of this paragraph: *Provided,* That:

(i) The package is designed such that it has a single "obvious principal display panel" as this term is defined in § 101.1 and has no other available surface area for an information panel or alternate principal display panel.

(ii) The area of surface available for labeling on the principal display panel of the package as this term is defined in § 101.1 is less than 12 square inches and bears all labeling appearing on the package.

(iii) The label information includes:

(a) Nutrition labeling in accordance with § 101.9.

(b) A full list of ingredients in accordance with regulations in this part and the policy expressed in § 101.6.

(iv) The information required by paragraph (b) of this section appears on the single, obvious principal display panel in accordance with the provisions of this paragraph (c) except that the type size is not less than one thirty-second inch in height.

(3) Packaged foods are exempt from the type size requirements of this paragraph: *Provided,* That:

(i) The package is designed such that it has a total surface area available to bear labeling of less than 12 square inches.

(ii) The label information includes:

(a) Nutrition labeling in accordance with § 101.9.

(b) A full list of ingredients in accordance with regulations in this part and the policy expressed in § 101.6.

(iii) The information required by paragraph (b) of this section appears on the principal display panel or information panel label in accordance with the provisions of this paragraph (c) except that the type size is not less than one thirty-second inch in height.

(4)(i) Soft drinks packaged in bottles manufactured before October 31, 1975 shall be exempt from the requirements prescribed by this section to the extent that information which is blown, lithographed, or formed onto the surface of the bottle is exempt from the size and placement requirements of this section.

(ii) Soft drinks packaged in bottles shall be exempt from the size and placement requirements prescribed by this section if all of the following conditions are met:

(a) If the soft drink is packaged in a bottle bearing a paper, plastic foam jacket, or foil label, or is packaged in a nonreusable bottle bearing a label lithographed onto the surface of the bottle or is packaged in metal cans, the product shall not be exempt from any requirement of this section other than the exemptions created by § 1.24(a)(5) (ii) and (v) of this chapter and the label shall bear all required information in the specified minimum type size, except the label will not be required to bear the information required by § 101.5 if this information appears on the bottle closure or on the lid of the can in a type size not less than one-sixteenth inch in height, or if embossed on the lid of the can in a type size not less than one-eighth inch in height.

(b) If the soft drink is packaged in a bottle which does not bear a paper,

plastic foam jacket or foil label, or is packaged in a reusable bottle bearing a label lithographed onto the surface of the bottle:

(1) Neither the bottle nor the closure is required to bear nutrition labeling in compliance with § 101.9, except that any multiunit retail package in which it is contained shall bear nutrition labeling if required by § 101.9; and any vending machine in which it is contained shall bear nutrition labeling if nutrition labeling is not present on the bottle or closure, if required by § 101.9.

(2) All other information pursuant to this section shall appear on the top of the bottle closure prominently and conspicuously in letters and/or numbers no less than one thirty-second inch in height, except that if the information required by § 101.5 is placed on the side of the closure in accordance with § 1.24(a)(5)(ii) of this chapter, such information shall appear in letters and/or numbers no less than one-sixteenth inch in height.

(3) Upon the petition of any interested person demonstrating that the bottle closure is too small to accommodate this information, the Commissioner may by regulation establish an alternative method of disseminating such information. Information appearing on the closure shall appear in the following priority:

(i) The warning required by § 100.130 of this chapter.

(ii) The statement of ingredients.

(iii) The name and address of the manufacturer, packer, or distributor.

(iv) The statement of identity.

(5) Individual serving-size packages of food served with meals in restaurants, institutions, and on board passenger carriers, and not intended for sale at retail, are exempt from type-size requirements of this paragraph, provided:

(i) The package has a total area of 3 square inches or less available to bear labeling;

(ii) There is insufficient area on the package available to print all required information in a type size of 1/16 inch in height;

(iii) The label information includes a full list of ingredients in accordance with regulations in this part and the

Food and Drug Administration, HHS

§ 101.2

the type size is not less than one thirty-second inch in height.

(3) Packaged foods are exempt from the type size requirements of this paragraph: *Provided,* That:

(i) The package is designed such that it has a total surface area available to bear labeling of less than 12 square inches.

(ii) The label information includes:

(a) Nutrition labeling in accordance with § 101.9.

(b) A full list of ingredients in accordance with regulations in this part and the policy expressed in § 101.6.

(iii) The information required by paragraph (b) of this section appears on the principal display panel or information panel label in accordance with the provisions of this paragraph (c) except that the type size is not less than one thirty-second inch in height.

(4)(i) Soft drinks packaged in bottles manufactured before October 31, 1975 shall be exempt from the requirements prescribed by this section to the extent that information which is blown, lithographed, or formed onto the surface of the bottle is exempt from the size and placement requirements of this section.

(ii) Soft drinks packaged in bottles shall be exempt from the size and placement requirements prescribed by this section if all of the following conditions are met:

(a) If the soft drink is packaged in a bottle bearing a paper, plastic foam jacket, or foil label, or is packaged in a nonreusable bottle bearing a label lithographed onto the surface of the bottle or is packaged in metal cans, the product shall not be exempt from any requirement of this section other than the exemptions created by § 1.24(a)(5) (ii) and (v) of this chapter and the label shall bear all required information in the specified minimum type size, except the label will not be required to bear the information required by § 101.5 if this information appears on the bottle closure or on the lid of the can in a type size not less than one-sixteenth inch in height, or if embossed on the lid of the can in a type size not less than one-eighth inch in height.

(b) If the soft drink is packaged in a bottle which does not bear a paper,

plastic foam jacket or foil label, or is packaged in a reusable bottle bearing a label lithographed onto the surface of the bottle:

(1) Neither the bottle nor the closure is required to bear nutrition labeling in compliance with § 101.9, except that any multiunit retail package in which it is contained shall bear nutrition labeling if required by § 101.9; and any vending machine in which it is contained shall bear nutrition labeling if nutrition labeling is not present on the bottle or closure, if required by § 101.9.

(2) All other information pursuant to this section shall appear on the top of the bottle closure prominently and conspicuously in letters and/or numbers no less than one thirty-second inch in height, except that if the information required by § 101.5 is placed on the side of the closure in accordance with § 1.24(a)(5)(ii) of this chapter, such information shall appear in letters and/or numbers no less than one-sixteenth inch in height.

(3) Upon the petition of any interested person demonstrating that the bottle closure is too small to accommodate this information, the Commissioner may by regulation establish an alternative method of disseminating such information. Information appearing on the closure shall appear in the following priority:

(i) The warning required by § 100.130 of this chapter.

(ii) The statement of ingredients.

(iii) The name and address of the manufacturer, packer, or distributor.

(iv) The statement of identity.

(5) Individual serving-size packages of food served with meals in restaurants, institutions, and on board passenger carriers, and not intended for sale at retail, are exempt from type-size requirements of this paragraph, provided:

(i) The package has a total area of 3 square inches or less available to bear labeling;

(ii) There is insufficient area on the package available to print all required information in a type size of ¹⁄₁₆ inch in height;

(iii) The label information includes a full list of ingredients in accordance with regulations in this part and the policy expressed in § 101.6 of this chapter; and

(iv) The information required by paragraph (b) of this section appears on the label in accordance with the provisions of this paragraph, except that the type size is not less than ½₂ inch in height.

(d)(1) All information required to appear on the principal display panel or on the information panel pursuant to this section shall appear on the same panel unless there is insufficient space. In determining the sufficiency of the available space, any vignettes, design, and other nonmandatory label information shall not be considered. If there is insufficient space for all of this information to appear on a single panel, it may be divided between these two panels except that the information required pursuant to any given section or part shall all appear on the same panel. A food whose label is required to bear the ingredient statement on the principal display panel may bear all other information specified in paragraph (b) of this section on the information panel.

(2) Any food, not otherwise exempted in this section, if packaged in a container consisting of a separate lid and body, and bearing nutrition labeling pursuant to § 101.9, and if the lid qualifies for and is designed to serve as a principal display panel, shall be exempt from the placement requirements of this section in the following respects:

(i) The name and place of business information required by § 101.5 shall not be required on the body of the container if this information appears on the lid in accordance with this section.

(ii) The nutrition information required by § 101.9 shall not be required on the lid if this information appears on the container body in accordance with this section.

(iii) The statement of ingredients required by § 101.4 shall not be required on the lid if this information appears on the container body in accordance with this section. Further, the statement of ingredients is not required on the container body if this information appears on the lid in accordance with this section.

(e) All information appearing on the information panel pursuant to this section shall appear in one place without other intervening material.

(f) If the label of any package of food is too small to accommodate all of the information required by §§ 101.4, 101.5, 101.8, 101.9, 101.17, and 101.25, and Part 105 of this chapter, the Commissioner may establish by regulation an acceptable alternative method of disseminating such information to the public, e.g., a type size smaller than one-sixteenth inch in height, or labeling attached to or inserted in the package or available at the point of purchase. A petition requesting such a regulation, as an amendment to this paragraph shall be submitted pursuant to Part 10 of this chapter.

[42 FR 14308, Mar. 15, 1977, as amended at 42 FR 15673, Mar. 22, 1977; 42 FR 45905, Sept. 13, 1977; 42 FR 47191, Sept. 20, 1977; 44 FR 16006, Mar. 16, 1979; 49 FR 13339, Apr. 4, 1984; 53 FR 16068, May 5, 1988]

§ 101.3 Identity labeling of food in packaged form.

(a) The principal display panel of a food in package form shall bear as one of its principal features a statement of the identity of the commodity.

(b) Such statement of identity shall be in terms of:

(1) The name now or hereafter specified in or required by any applicable Federal law or regulation; or, in the absence thereof,

(2) The common or usual name of the food; or, in the absence thereof,

(3) An appropriately descriptive term, or when the nature of the food is obvious, a fanciful name commonly used by the public for such food.

(c) Where a food is marketed in various optional forms (whole, slices, diced, etc.), the particular form shall be considered to be a necessary part of the statement of identity and shall be declared in letters of a type size bearing a reasonable relation to the size of the letters forming the other components of the statement of identity; except that if the optional form is visible through the container or is depicted by an appropriate vignette, the particular form need not be included in the statement. This specification does not affect the required declarations of identity under definitions and standards for foods promulgated pursuant to section 401 of the act.

(d) This statement of identity shall be presented in bold type on the principal display panel, shall be in a size reasonably related to the most prominent printed matter on such panel, and shall be in lines generally parallel to the base on which the package rests as it is designed to be displayed.

(e) Under the provisions of section 403(c) of the Federal Food, Drug, and Cosmetic Act, a food shall be deemed to be misbranded if it is an imitation of another food unless its label bears, in type of uniform size and prominence, the word "imitation" and, immediately thereafter, the name of the food imitated.

(1) A food shall be deemed to be an imitation and thus subject to the requirements of section 403(c) of the act if it is a substitute for and resembles another food but is nutritionally inferior to that food.

(2) A food that is a substitute for and resembles another food shall not be deemed to be an imitation provided it meets each of the following requirements:

(i) It is not nutritionally inferior to the food for which it substitutes and which it resembles.

(ii) Its label bears a common or usual name that complies with the provisions of § 102.5 of this chapter and that is not false or misleading, or in the absence of an existing common or usual name, an appropriately descriptive term that is not false or misleading. The label may, in addition, bear a fanciful name which is not false or misleading.

(3) A food for which a common or usual name is established by regulation (e.g., in a standard of identity pursuant to section 401 of the act, in a common or usual name regulation pursuant to Part 102 of this chapter, or in a regulation establishing a nutritional quality guideline pursuant to Part 104 of this chapter), and which complies with all of the applicable requirements of such regulation(s), shall not be deemed to be an imitation.

(4) Nutritional inferiority includes:

(i) Any reduction in the content of an essential nutrient that is present in a measurable amount, but does not include a reduction in the caloric or fat content provided the food is labeled pursuant to the provisions of § 101.9, and provided the labeling with respect to any reduction in caloric content complies with the provisions applicable to caloric content in Part 105 of this chapter.

(ii) For the purpose of this section, a measurable amount of an essential nutrient in a food shall be considered to be 2 percent or more of the U.S. RDA of protein or any vitamin or mineral listed under § 101.9(c)(7)(iv) of this chapter per average or usual serving, or where the food is customarily not consumed directly, per average or usual portion, as established in § 101.9.

(iii) If the Commissioner concludes that a food is a substitute for and resembles another food but is inferior to the food imitated for reasons other than those set forth in this paragraph, he may propose appropriate revisions to this regulation or he may propose a separate regulation governing the particular food.

(f) A label may be required to bear the percentage(s) of a characterizing ingredient(s) or information concerning the presence or absence of an ingredient(s) or the need to add an ingredient(s) as part of the common or usual name of the food pursuant to Subpart B of Part 102 of this chapter.

[42 FR 14308, Mar. 15, 1977, as amended at 48 FR 10811, Mar. 15, 1983]

§ 101.4 Food; designation of ingredients.

(a) Ingredients required to be declared on the label of a food, including foods that comply with standards of identity that require labeling in compliance with this Part 101, except those exempted by § 101.100, shall be listed by common or usual name in descending order of predominance by weight on either the principal display panel or the information panel in accordance with the provisions of § 101.2.

(b) The name of an ingredient shall be a specific name and not a collective (generic) name, except that:

(1) Spices, flavorings, colorings and chemical preservatives shall be declared according to the provisions of § 101.22.

(2) An ingredient which itself contains two or more ingredients and which has an established common or usual name, conforms to a standard established pursuant to the Meat Inspection or Poultry Products Inspection Acts by the U.S. Department of Agriculture, or conforms to a definition and standard of identity established pursuant to section 401 of the Federal Food, Drug, and Cosmetic Act, shall be designated in the statement of ingredients on the label of such food by either of the following alternatives:

(i) By declaring the established common or usual name of the ingredient followed by a parenthetical listing of all ingredients contained therein in descending order of predominance except that, if the ingredient is a food subject to a definition and standard of identity established in this Subchapter B, only the ingredients required to be declared by the definition and standard of identity need be listed; or

(ii) By incorporating into the statement of ingredients in descending order of predominance in the finished food, the common or usual name of every component of the ingredient without listing the ingredient itself.

(3) Skim milk, concentrated skim milk, reconstituted skim milk, and nonfat dry milk may be declared as "skim milk" or "nonfat milk".

(4) Milk, concentrated milk, reconstituted milk, and dry whole milk may be declared as "milk".

(5) Bacterial cultures may be declared by the word "cultured" followed by the name of the substrate, e.g., "made from cultured skim milk or cultured buttermilk".

(6) Sweetcream buttermilk, concentrated sweetcream buttermilk, reconstituted sweetcream buttermilk, and dried sweetcream buttermilk may be declared as "buttermilk".

(7) Whey, concentrated whey, reconstituted whey, and dried whey may be declared as "whey".

(8) Cream, reconstituted cream, dried cream, and plastic cream (sometimes known as concentrated milk fat) may be declared as "cream".

(9) Butteroil and anhydrous butterfat may be declared as "butterfat".

(10) Dried whole eggs, frozen whole eggs, and liquid whole eggs may be declared as "eggs".

(11) Dried egg whites, frozen egg whites, and liquid egg whites may be declared as "egg whites".

(12) Dried egg yolks, frozen egg yolks, and liquid egg yolks may be declared as "egg yolks".

(13) [Reserved]

(14) Each individual fat and/or oil ingredient of a food intended for human consumption shall be declared by its specific common or usual name (e.g., "beef fat", "cottonseed oil") in its order of predominance in the food except that blends of fats and/or oils may be designated in their order of predominance in the foods as "—— shortening" or "blend of —— oils", the blank to be filled in with the word "vegetable", "animal", "marine", with or without the terms "fat" or "oils", or combination of these, whichever is applicable if, immediately following the term, the common or usual name of each individual vegetable, animal, or marine fat or oil is given in parentheses, e.g., "vegetable oil shortening (soybean and cottonseed oil)". For products that are blends of fats and/or oils and for foods in which fats and/or oils constitute the predominant ingredient, i.e., in which the combined weight of all fat and/or oil ingredients equals or exceeds the weight of the most predominant ingredient that is not a fat or oil, the listing of the common or usual names of such fats and/or oils in parentheses shall be in descending order of predominance. In all other foods in which a blend of fats and/or oils is used as an ingredient, the listing of the common or usual names in parentheses need not be in descending order of predominance if the manufacturer, because of the use of varying mixtures, is unable to adhere to a constant pattern of fats and/or oils in the product. If the fat or oil is completely hydrogenated, the name shall include the term "hydrogenated", or if partially hydrogenated, the name shall include the term "partially hydrogenated". If each fat and/or oil in a blend or the blend is completely hydrogenated, the term "hy-

Food and Drug Administration, HHS

§ 101.4

drogenated" may precede the term(s) describing the blend, e.g., "hydrogenated vegetable oil (soybean, cottonseed, and palm oils)", rather than preceding the name of each individual fat and/or oil; if the blend of fats and/or oils is partially hydrogenated, the term "partially hydrogenated" may be used in the same manner. Fat and/or oil ingredients not present in the product may be listed if they may sometimes be used in the product. Such ingredients shall be identified by words indicating that they may not be present, such as "or", "and/or", "contains one or more of the following:", e.g., "vegetable oil shortening (contains one or more of the following: cottonseed oil, palm oil, soybean oil)". No fat or oil ingredient shall be listed unless actually present if the fats and/or oils constitute the predominant ingredient of the product, as defined in this paragraph (b)(14).

(15) When all the ingredients of a wheat flour are declared in an ingredient statement, the principal ingredient of the flour shall be declared by the name(s) specified in §§ 137.105, 137.200, 137.220 and 137.225 of this chapter, i.e., the first ingredient designated in the ingredient list of flour, or bromated flour, or enriched flour, or self-rising flour is "flour", "white flour", "wheat flour", or "plain flour"; the first ingredient designated in the ingredient list of durum flour is "durum flour"; the first ingredient designated in the ingredient list of whole wheat flour, or bromated whole wheat flour is "whole wheat flour", "graham flour", or "entire wheat flour"; and the first ingredient designated in the ingredient list of whole durum wheat flour is "whole durum wheat flour".

(16) Ingredients that act as leavening agents in food may be declared in the ingredient statement by stating the specific common or usual name of each individual leavening agent in parentheses following the collective name "leavening", e.g., "leavening (baking soda, monocalcium phosphate, and calcium carbonate)". The listing of the common or usual name of each individual leavening agent in parentheses shall be in descending order of predominance: *Except*, That if the

manufacturer is unable to adhere to a constant pattern of leavening agents in the product, the listing of individual leavening agents need not be in descending order of predominance. Leavening agents not present in the product may be listed if they are sometimes used in the product. Such ingredients shall be identified by words indicating that they may not be present, such as "or", "and/or", "contains one or more of the following:".

(17) Ingredients that act as yeast nutrients in foods may be declared in the ingredient statement by stating the specific common or usual name of each individual yeast nutrient in parentheses following the collective name "yeast nutrients", e.g., "yeast nutrients (calcium sulfate and ammonium phosphate)". The listing of the common or usual name of each individual yeast nutrient in parentheses shall be in descending order of predominance: *Except*, That if the manufacturer is unable to adhere to a constant pattern of yeast nutrients in the product, the listing of the common or usual names of individual yeast nutrients need not be in descending order of predominance. Yeast nutrients not present in the product may be listed if they are sometimes used in the product. Such ingredients shall be identified by words indicating that they may not be present, such as "or", "and/or", or "contains one or more of the following:".

(18) Ingredients that act as dough conditioners may be declared in the ingredient statement by stating the specific common or usual name of each individual dough conditioner in parentheses following the collective name "dough conditioner", e.g., "dough conditioners (L-cysteine, ammonium sulfate)". The listing of the common or usual name of each dough conditioner in parentheses shall be in descending order of predominance: *Except*, That if the manufacturer is unable to adhere to a constant pattern of dough conditioners in the product, the listing of the common or usual names of individual dough conditioners need not be in descending order of predominance. Dough conditioners not present in the product may be listed if they are sometimes used in the product. Such

§ 101.5   **21 CFR Ch. I (4-1-90 Edition)**

ingredients shall be identified by words indicating that they may not be present, such as "or", "and/or", or "contains one or more of the following:".

(19) Ingredients that act as firming agents in food (e.g., salts of calcium and other safe and suitable salts in canned vegetables) may be declared in the ingredient statement, in order of predominance appropriate for the total of all firming agents in the food, by stating the specific common or usual name of each individual firming agent in descending order of predominance in parentheses following the collective name "firming agents". If the manufacturer is unable to adhere to a constant pattern of firming agents in the food, the listing of the individual firming agents need not be in descending order of predominance. Firming agents not present in the product may be listed if they are sometimes used in the product. Such ingredients shall be identified by words indicating that they may not be present, such as "or", "and/or", "contains one or more of the following:".

(c) When water is added to reconstitute, completely or partially, an ingredient permitted by paragraph (b) of this section to be declared by a class name, the position of the ingredient class name in the ingredient statement shall be determined by the weight of the unreconstituted ingredient plus the weight of the quantity of water added to reconstitute that ingredient, up to the amount of water needed to reconstitute the ingredient to single strength. Any water added in excess of the amount of water needed to reconstitute the ingredient to single strength shall be declared as "water" in the ingredient statement.

[42 FR 14308, Mar. 15, 1977, as amended at 43 FR 12858, Mar. 28, 1978; 43 FR 24519, June 6, 1978; 48 FR 8054, Feb. 25, 1983]

§ 101.5 **Food; name and place of business of manufacturer, packer, or distributor.**

(a) The label of a food in packaged form shall specify conspicuously the name and place of business of the manufacturer, packer, or distributor.

(b) The requirement for declaration of the name of the manufacturer,

packer, or distributor shall be deemed to be satisfied, in the case of a corporation, only by the actual corporate name, which may be preceded or followed by the name of the particular division of the corporation. In the case of an individual, partnership, or association, the name under which the business is conducted shall be used.

(c) Where the food is not manufactured by the person whose name appears on the label, the name shall be qualified by a phrase that reveals the connection such person has with such food; such as "Manufactured for _____", "Distributed by _____", or any other wording that expresses the facts.

(d) The statement of the place of business shall include the street address, city, State, and ZIP code; however, the street address may be omitted if it is shown in a current city directory or telephone directory. The requirement for inclusion of the ZIP code shall apply only to consumer commodity labels developed or revised after the effective date of this section. In the case of nonconsumer packages, the ZIP code shall appear either on the label or the labeling (including invoice).

(e) If a person manufactures, packs, or distributes a food at a place other than his principal place of business, the label may state the principal place of business in lieu of the actual place where such food was manufactured or packed or is to be distributed, unless such statement would be misleading.

Part 110

Subpart C—Naturally Occurring Po-
sionous or Deleterious Substances [Reserved]

PART 110—CURRENT GOOD MANU-
FACTURING PRACTICE IN MANU-
FACTURING, PACKING, OR HOLD-
ING HUMAN FOOD

Subpart A—General Provisions

Sec.
110.3   Definitions.
110.5   Current good manufacturing practice.
110.10   Personnel.
110.19   Exclusions.

Subpart B—Buildings and Facilities

110.20   Plant and grounds.
110.35   Sanitary operations.
110.37   Sanitary facilities and controls.

Subpart C—Equipment

110.40   Equipment and utensils.

Subpart D—[Reserved]

Subpart E—Production and Process Controls

110.80   Processes and controls.
110.93   Warehousing and distribution.

Subpart F—[Reserved]

Subpart G—Defect Action Levels

110.110   Natural or unavoidable defects in
food for human use that present no
health hazard.

AUTHORITY: Secs. 402, 701, 704 of the Fed-
eral Food, Drug, and Cosmetic Act (21
U.S.C. 342, 371, 374); sec. 361 of the Public
Health Service Act (42 U.S.C. 264).

SOURCE: 51 FR 24475, June 19, 1986, unless
otherwise noted.

Subpart A—General Provisions

§ 110.3   Definitions.

The definitions and interpretations
of terms in section 201 of the Federal
Food, Drug, and Cosmetic Act (the
act) are applicable to such terms when
used in this part. The following definitions shall also apply:

(a) "Acid foods or acidified foods"
means foods that have an equilibrium
pH of 4.6 or below.

(b) "Adequate" means that which is
needed to accomplish the intended
purpose in keeping with good public
health practice.

(c) "Batter" means a semifluid substance, usually composed of flour and
other ingredients, into which principal
components of food are dipped or with
which they are coated, or which may
be used directly to form bakery foods.

(d) "Blanching," except for tree nuts
and peanuts, means a prepackaging
heat treatment of foodstuffs for a sufficient time and at a sufficient temperature to partially or completely inacti-
vate the naturally occurring enzymes
and to effect other physical or bio-
chemical changes in the food.

(e) "Critical control point" means a
point in a food process where there is
a high probability that improper control may cause, allow, or contribute to
a hazard or to filth in the final food or
decomposition of the final food.

(f) "Food" means food as defined in
section 201(f) of the act and includes
raw materials and ingredients.

(g) "Food-contact surfaces" are
those surfaces that contact human
food and those surfaces from which
drainage onto the food or onto surfaces that contact the food ordinarily
occurs during the normal course of operations. "Food-contact surfaces" includes utensils and food-contact surfaces of equipment.

(h) "Lot" means the food produced
during a period of time indicated by a
specific code.

(i) "Microorganisms" means yeasts,
molds, bacteria, and viruses and includes, but is not limited to, species
having public health significance. The
term "undesirable microorganisms" includes those microorganisms that are
of public health significance, that subject food to decomposition, that indicate that food is contaminated with
filth, or that otherwise may cause food
to be adulterated within the meaning
of the act. Occasionally in these regulations, FDA used the adjective "microbial" instead of using an adjectival
phrase containing the word microorga-
nism.

(j) "Pest" refers to any objectionable
animals or insects including, but not
limited to, birds, rodents, flies, and
larvae.

Food and Drug Administration, HHS   § 110.10

(k) "Plant" means the building or facility or parts thereof, used for or in connection with the manufacturing, packaging, labeling, or holding of human food.

(l) "Quality control operation" means a planned and systematic procedure for taking all actions necessary to prevent food from being adulterated within the meaning of the act.

(m) "Rework" means clean, unadulterated food that has been removed from processing for reasons other than insanitary conditions or that has been successfully reconditioned by reprocessing and that is suitable for use as food.

(n) "Safe-moisture level" is a level of moisture low enough to prevent the growth of undesirable microorganisms in the finished product under the intended conditions of manufacturing, storage, and distribution. The maximum safe moisture level for a food is based on its water activity (a_w). An a_w will be considered safe for a food if adequate data are available that demonstrate that the food at or below the given a_w will not support the growth of undesirable microorganisms.

(o) "Sanitize" means to adequately treat food-contact surfaces by a process that is effective in destroying vegetative cells of microorganisms of public health significance, and in substantially reducing numbers of other undesirable microorganisms, but without adversely affecting the product or its safety for the consumer.

(p) "Shall" is used to state mandatory requirements.

(q) "Should" is used to state recommended or advisory procedures or identify recommended equipment.

(r) "Water activity" (a_w) is a measure of the free moisture in a food and is the quotient of the water vapor pressure of the substance divided by the vapor pressure of pure water at the same temperature.

§ 110.5 Current good manufacturing practice.

(a) The criteria and definitions in this part shall apply in determining whether a food is adulterated (1) within the meaning of section 402(a)(3) of the act in that the food has been manufactured under such conditions that it is unfit for food; or (2) within the meaning of section 402(a)(4) of the act in that the food has been prepared, packed, or held under insanitary conditions whereby it may have become contaminated with filth, or whereby it may have been rendered injurious to health. The criteria and definitions in this part also apply in determining whether a food is in violation of section 361 of the Public Health Service Act (42 U.S.C. 264).

(b) Food covered by specific current good manufacturing practice regulations also is subject to the requirements of those regulations.

§ 110.10 Personnel.

The plant management shall take all reasonable measures and precautions to ensure the following:

(a) *Disease control.* Any person who, by medical examination or supervisory observation, is shown to have, or appears to have, an illness, open lesion, including boils, sores, or infected wounds, or any other abnormal source of microbial contamination by which there is a reasonable possibility of food, food-contact surfaces, or food-packaging materials becoming contaminated, shall be excluded from any operations which may be expected to result in such contamination until the condition is corrected. Personnel shall be instructed to report such health conditions to their supervisors.

(b) *Cleanliness.* All persons working in direct contact with food, food-contact surfaces, and food-packaging materials shall conform to hygienic practices while on duty to the extent necessary to protect against contamination of food. The methods for maintaining cleanliness include, but are not limited to:

(1) Wearing outer garments suitable to the operation in a manner that protects against the contamination of food, food-contact surfaces, or food-packaging materials.

(2) Maintaining adequate personal cleanliness.

(3) Washing hands thoroughly (and sanitizing if necessary to protect against contamination with undesirable microorganisms) in an adequate

§ 110.19

21 CFR Ch. I (4-1-90 Edition)

hand-washing facility before starting work, after each absence from the work station, and at any other time when the hands may have become soiled or contaminated.

(4) Removing all unsecured jewelry and other objects that might fall into food, equipment, or containers, and removing hand jewelry that cannot be adequately sanitized during periods in which food is manipulated by hand. If such hand jewelry cannot be removed, it may be covered by material which can be maintained in an intact, clean, and sanitary condition and which effectively protects against the contamination by these objects of the food, food-contact surfaces, or food-packaging materials.

(5) Maintaining gloves, if they are used in food handling, in an intact, clean, and sanitary condition. The gloves should be of an impermeable material.

(6) Wearing, where appropriate, in an effective manner, hair nets, headbands, caps, beard covers, or other effective hair restraints.

(7) Storing clothing or other personal belongings in areas other than where food is exposed or where equipment or utensils are washed.

(8) Confining the following to areas other than where food may be exposed or where equipment or utensils are washed: eating food, chewing gum, drinking beverages, or using tobacco.

(9) Taking any other necessary precautions to protect against contamination of food, food-contact surfaces, or food-packaging materials with microorganisms or foreign substances including, but not limited to, perspiration, hair, cosmetics, tobacco, chemicals, and medicines applied to the skin.

(c) Education and training. Personnel responsible for identifying sanitation failures or food contamination should have a background of education or experience, or a combination thereof, to provide a level of competency necessary for production of clean and safe food. Food handlers and supervisors should receive appropriate training in proper food handling techniques and food-protection principles and should be informed of the danger of poor personal hygiene and insanitary practices.

(d) Supervision. Responsibility for assuring compliance by all personnel with all requirements of this part shall be clearly assigned to competent supervisory personnel.

[51 FR 24475, June 19, 1986, as amended at 54 FR 24892, June 12, 1989]

§ 110.19 Exclusions.

(a) The following operations are not subject to this part: Establishments engaged solely in the harvesting, storage, or distribution of one or more "raw agricultural commodities," as defined in section 201(r) of the act, which are ordinarily cleaned, prepared, treated, or otherwise processed before being marketed to the consuming public.

(b) FDA, however, will issue special regulations if it is necessary to cover these excluded operations.

Subpart B—Buildings and Facilities

§ 110.20 Plant and grounds.

(a) Grounds. The grounds about a food plant under the control of the operator shall be kept in a condition that will protect against the contamination of food. The methods for adequate maintenance of grounds include, but are not limited to:

(1) Properly storing equipment, removing litter and waste, and cutting weeds or grass within the immediate vicinity of the plant buildings or structures that may constitute an attractant, breeding place, or harborage for pests.

(2) Maintaining roads, yards, and parking lots so that they do not constitute a source of contamination in areas where food is exposed.

(3) Adequately draining areas that may contribute contamination to food by seepage, foot-borne filth, or providing a breeding place for pests.

(4) Operating systems for waste treatment and disposal in an adequate manner so that they do not constitute a source of contamination in areas where food is exposed.

If the plant grounds are bordered by grounds not under the operator's control and not maintained in the manner described in paragraph (a) (1) through (3) of this section, care shall be exercised in the plant by inspection, extermination, or other means to exclude pests, dirt, and filth that may be a source of food contamination.

(b) *Plant construction and design.* Plant buildings and structures shall be suitable in size, construction, and design to facilitate maintenance and sanitary operations for food-manufacturing purposes. The plant and facilities shall:

(1) Provide sufficient space for such placement of equip- ment and storage of materials as is necessary for the maintenance of sanitary operations and the production of safe food.

(2) Permit the taking of proper precautions to reduce the potential for contamination of food, food-contact surfaces, or food-packaging materials with microorganisms, chemicals, filth, or other extraneous material. The potential for contamination may be reduced by adequate food safety controls and operating practices or effective design, including the separation of operations in which contamination is likely to occur, by one or more of the following means: location, time, partition, air flow, enclosed systems, or other effective means.

(3) Permit the taking of proper precautions to protect food in outdoor bulk fermentation vessels by any effective means, including:

(i) Using protective coverings.

(ii) Controlling areas over and around the vessels to eliminate harborages for pests.

(iii) Checking on a regular basis for pests and pest infestation.

(iv) Skimming the fermentation vessels, as necessary.

(4) Be constructed in such a manner that floors, walls, and ceilings may be adequately cleaned and kept clean and kept in good repair; that drip or condensate from fixtures, ducts and pipes does not contaminate food, food-contact surfaces, or food-packaging materials; and that aisles or working spaces are provided between equipment and walls and are adequately unobstructed and of adequate width to permit employees to perform their duties and to protect against contaminating food or food-contact surfaces with clothing or personal contact.

(5) Provide adequate lighting in hand-washing areas, dressing and locker rooms, and toilet rooms and in all areas where food is examined, processed, or stored and where equipment or utensils are cleaned; and provide safety-type light bulbs, fixtures, skylights, or other glass suspended over exposed food in any step of preparation or otherwise protect against food contamination in case of glass breakage.

(6) Provide adequate ventilation or control equipment to minimize odors and vapors (including steam and noxious fumes) in areas where they may contaminate food; and locate and operate fans and other air-blowing equipment in a manner that minimizes the potential for contaminating food, food-packaging materials, and food-contact surfaces.

(7) Provide, where necessary, adequate screening or other protection against pests.

§ 110.35 Sanitary operations.

(a) *General maintenance.* Buildings, fixtures, and other physical facilities of the plant shall be maintained in a sanitary condition and shall be kept in repair sufficient to prevent food from becoming adulterated within the meaning of the act. Cleaning and sanitizing of utensils and equipment shall be conducted in a manner that protects against contamination of food, food-contact surfaces, or food-packaging materials.

(b) *Substances used in cleaning and sanitizing; storage of toxic materials.* (1) Cleaning compounds and sanitizing agents used in cleaning and sanitizing procedures shall be free from undesirable microorganisms and shall be safe and adequate under the conditions of use. Compliance with this requirement may be verified by any effective means including purchase of these substances under a supplier's guarantee or certification, or examination of these substances for contamination. Only the following toxic materials may be used or stored in a plant where food is processed or exposed:

(i) Those required to maintain clean and sanitary conditions;

§ 110.37

21 CFR Ch. I (4-1-90 Edition)

(ii) Those necessary for use in laboratory testing procedures;

(iii) Those necessary for plant and equipment maintenance and operation; and

(iv) Those necessary for use in the plant's operations.

(2) Toxic cleaning compounds, sanitizing agents, and pesticide chemicals shall be identified, held, and stored in a manner that protects against contamination of food, food-contact surfaces, or food-packaging materials. All relevant regulations promulgated by other Federal, State, and local government agencies for the application, use, or holding of these products should be followed.

(c) *Pest control.* No pests shall be allowed in any area of a food plant. Guard or guide dogs may be allowed in some areas of a plant if the presence of the dogs is unlikely to result in contamination of food, food-contact surfaces, or food-packaging materials. Effective measures shall be taken to exclude pests from the processing areas and to protect against the contamination of food on the premises by pests. The use of insecticides or rodenticides is permitted only under precautions and restrictions that will protect against the contamination of food, food-contact surfaces, and food-packaging materials.

(d) *Sanitation of food-contact surfaces.* All food-contact surfaces, including utensils and food-contact surfaces of equipment, shall be cleaned as frequently as necessary to protect against contamination of food.

(1) Food-contact surfaces used for manufacturing or holding low-moisture food shall be in a dry, sanitary condition at the time of use. When the surfaces are wet-cleaned, they shall, when necessary, be sanitized and thoroughly dried before subsequent use.

(2) In wet processing, when cleaning is necessary to protect against the introduction of microorganisms into food, all food-contact surfaces shall be cleaned and sanitized before use and after any interruption during which the food-contact surfaces may have become contaminated. Where equipment and utensils are used in a continuous production operation, the utensils and food-contact surfaces of the

equipment shall be cleaned and sanitized as necessary.

(3) Non-food-contact surfaces of equipment used in the operation of food plants should be cleaned as frequently as necessary to protect against contamination of food.

(4) Single-service articles (such as utensils intended for one-time use, paper cups, and paper towels) should be stored in appropriate containers and shall be handled, dispensed, used, and disposed of in a manner that protects against contamination of food or food-contact surfaces.

(5) Sanitizing agents shall be adequate and safe under conditions of use. Any facility, procedure, or machine is acceptable for cleaning and sanitizing equipment and utensils if it is established that the facility, procedure, or machine will routinely render equipment and utensils clean and provide adequate cleaning and sanitizing treatment.

(e) *Storage and handling of cleaned portable equipment and utensils.* Cleaned and sanitized portable equipment with food-contact surfaces and utensils should be stored in a location and manner that protects food-contact surfaces from contamination.

[51 FR 24475, June 19, 1986, as amended at 54 FR 24892, June 12, 1989]

§ 110.37 Sanitary facilities and controls.

Each plant shall be equipped with adequate sanitary facilities and accommodations including, but not limited to:

(a) *Water supply.* The water supply shall be sufficient for the operations intended and shall be derived from an adequate source. Any water that contacts food or food-contact surfaces shall be safe and of adequate sanitary quality. Running water at a suitable temperature, and under pressure as needed, shall be provided in all areas where required for the processing of food, for the cleaning of equipment, utensils, and food-packaging materials, or for employee sanitary facilities.

(b) *Plumbing.* Plumbing shall be of adequate size and design and adequately installed and maintained to:

Food and Drug Administration, HHS   § 110.40

(1) Carry sufficient quantities of water to required locations throughout the plant.

(2) Properly convey sewage and liquid disposable waste from the plant.

(3) Avoid constituting a source of contamination to food, water supplies, equipment, or utensils or creating an unsanitary condition.

(4) Provide adequate floor drainage in all areas where floors are subject to flooding-type cleaning or where normal operations release or discharge water or other liquid waste on the floor.

(5) Provide that there is not backflow from, or cross-connection between, piping systems that discharge waste water or sewage and piping systems that carry water for food or food manufacturing.

(c) *Sewage disposal.* Sewage disposal shall be made into an adequate sewerage system or disposed of through other adequate means.

(d) *Toilet facilities.* Each plant shall provide its employees with adequate, readily accessible toilet facilities. Compliance with this requirement may be accomplished by:

(1) Maintaining the facilities in a sanitary condition.

(2) Keeping the facilities in good repair at all times.

(3) Providing self-closing doors.

(4) Providing doors that do not open into areas where food is exposed to airborne contamination, except where alternate means have been taken to protect against such contamination (such as double doors or positive airflow systems).

(e) *Hand-washing facilities.* Hand-washing facilities shall be adequate and convenient and be furnished with running water at a suitable temperature. Compliance with this requirement may be accomplished by providing:

(1) Hand-washing and, where appropriate, hand-sanitizing facilities at each location in the plant where good sanitary practices require employees to wash and/or sanitize their hands.

(2) Effective hand-cleaning and sanitizing preparations.

(3) Sanitary towel service or suitable drying devices.

(4) Devices or fixtures, such as water control valves, so designed and constructed to protect against recontamination of clean, sanitized hands.

(5) Readily understandable signs directing employees handling unprotected food, unprotected food-packaging materials, of food-contact surfaces to wash and, where appropriate, sanitize their hands before they start work, after each absence from post of duty, and when their hands may have become soiled or contaminated. These signs may be posted in the processing room(s) and in all other areas where employees may handle such food, materials, or surfaces.

(6) Refuse receptacles that are constructed and maintained in a manner that protects against contamination of food.

(f) *Rubbish and offal disposal.* Rubbish and any offal shall be so conveyed, stored, and disposed of as to minimize the development of odor, minimize the potential for the waste becoming an attractant and harborage or breeding place for pests, and protect against contamination of food, food-contact surfaces, water supplies, and ground surfaces.

Subpart C—Equipment

§ 110.40   Equipment and utensils.

(a) All plant equipment and utensils shall be so designed and of such material and workmanship as to be adequately cleanable, and shall be properly maintained. The design, construction, and use of equipment and utensils shall preclude the adulteration of food with lubricants, fuel, metal fragments, contaminated water, or any other contaminants. All equipment should be so installed and maintained as to facilitate the cleaning of the equipment and of all adjacent spaces. Food-contact surfaces shall be corrosion-resistant when in contact with food. They shall be made of nontoxic materials and designed to withstand the environment of their intended use and the action of food, and, if applicable, cleaning compounds and sanitizing agents. Food-contact surfaces shall be maintained to protect food from being

§ 110.80

21 CFR Ch. I (4-1-90 Edition)

contaminated by any source, including unlawful indirect food additives.

(b) Seams on food-contact surfaces shall be smoothly bonded or maintained so as to minimize accumulation of food particles, dirt, and organic matter and thus minimize the opportunity for growth of microorganisms.

(c) Equipment that is in the manufacturing or food-handling area and that does not come into contact with food shall be so constructed that it can be kept in a clean condition.

(d) Holding, conveying, and manufacturing systems, including gravimetric, pneumatic, closed, and automated systems, shall be of a design and construction that enables them to be maintained in an appropriate sanitary condition.

(e) Each freezer and cold storage compartment used to store and hold food capable of supporting growth of microorganisms shall be fitted with an indicating thermometer, temperature-measuring device, or temperature-recording device so installed as to show the temperature accurately within the compartment, and should be fitted with an automatic control for regulating temperature or with an automatic alarm system to indicate a significant temperature change in a manual operation.

(f) Instruments and controls used for measuring, regulating, or recording temperatures, pH, acidity, water activity, or other conditions that control or prevent the growth of undesirable microorganisms in food shall be accurate and adequately maintained, and adequate in number for their designated uses.

(g) Compressed air or other gases mechanically introduced into food or used to clean food-contact surfaces or equipment shall be treated in such a way that food is not contaminated with unlawful indirect food additives.

Subpart D—[Reserved]

Subpart E—Production and Process Controls

§ 110.80 Processes and controls.

All operations in the receiving, inspecting, transporting, segregating, preparing, manufacturing, packaging,

and storing of food shall be conducted in accordance with adequate sanitation principles. Appropriate quality control operations shall be employed to ensure that food is suitable for human consumption and that food-packaging materials are safe and suitable. Overall sanitation of the plant shall be under the supervision of one or more competent individuals assigned responsibility for this function. All reasonable precautions shall be taken to ensure that production procedures do not contribute contamination from any source. Chemical, microbial, or extraneous-material testing procedures shall be used where necessary to identify sanitation failures or possible food contamination. All food that has become contaminated to the extent that it is adulterated within the meaning of the act shall be rejected, or if permissible, treated or processed to eliminate the contamination.

(a) *Raw materials and other ingredients.* (1) Raw materials and other ingredients shall be inspected and segregated or otherwise handled as necessary to ascertain that they are clean and suitable for processing into food and shall be stored under conditions that will protect against contamination and minimize deterioration. Raw materials shall be washed or cleaned as necessary to remove soil or other contamination. Water used for washing, rinsing, or conveying food shall be safe and of adequate sanitary quality. Water may be reused for washing, rinsing, or conveying food if it does not increase the level of contamination of the food. Containers and carriers of raw materials should be inspected on receipt to ensure that their condition has not contributed to the contamination or deterioration of food.

(2) Raw materials and other ingredients shall either not contain levels of microorganisms that may produce food poisoning or other disease in humans, or they shall be pasteurized or otherwise treated during manufacturing operations so that they no longer contain levels that would cause the product to be adulterated within the meaning of the act. Compliance with this requirement may be verified by any effective means, including purchasing raw materials and other ingre-

dients under a supplier's guarantee or certification.

(3) Raw materials and other ingredients susceptible to contamination with aflatoxin or other natural toxins shall comply with current Food and Drug Administration regulations, guidelines, and action levels for poisonous or deleterious substances before these materials or ingredients are incorporated into finished food. Compliance with this requirement may be accomplished by purchasing raw materials and other ingredients under a supplier's guarantee or certification, or may be verified by analyzing these materials and ingredients for aflatoxins and other natural toxins.

(4) Raw materials, other ingredients, and rework susceptible to contamination with pests, undesirable microorganisms, or extraneous material shall comply with applicable Food and Drug Administration regulations, guidelines, and defect action levels for natural or unavoidable defects if a manufacturer wishes to use the materials in manufacturing food. Compliance with this requirement may be verified by any effective means, including purchasing the materials under a supplier's guarantee or certification, or examination of these materials for contamination.

(5) Raw materials, other ingredients, and rework shall be held in bulk, or in containers designed and constructed so as to protect against contamination and shall be held at such temperature and relative humidity and in such a manner as to prevent the food from becoming adulterated within the meaning of the act. Material scheduled for rework shall be identified as such.

(6) Frozen raw materials and other ingredients shall be kept frozen. If thawing is required prior to use, it shall be done in a manner that prevents the raw materials and other ingredients from becoming adulterated within the meaning of the act.

(7) Liquid or dry raw materials and other ingredients received and stored in bulk form shall be held in a manner that protects against contamination.

(b) *Manufacturing operations.* (1) Equipment and utensils and finished food containers shall be maintained in an acceptable condition through ap-

propriate cleaning and sanitizing, as necessary. Insofar as necessary, equipment shall be taken apart for thorough cleaning.

(2) All food manufacturing, including packaging and storage, shall be conducted under such conditions and controls as are necessary to minimize the potential for the growth of microorganisms, or for the contamination of food. One way to comply with this requirement is careful monitoring of physical factors such as time, temperature, humidity, a_w, pH, pressure, flow rate, and manufacturing operations such as freezing, dehydration, heat processing, acidification, and refrigeration to ensure that mechanical breakdowns, time delays, temperature fluctuations, and other factors do not contribute to the decomposition or contamination of food.

(3) Food that can support the rapid growth of undesirable microorganisms, particularly those of public health significance, shall be held in a manner that prevents the food from becoming adulterated within the meaning of the act. Compliance with this requirement may be accomplished by any effective means, including:

(i) Maintaining refrigerated foods at 45 °F (7.2 °C) or below as appropriate for the particular food involved.

(ii) Maintaining frozen foods in a frozen state.

(iii) Maintaining hot foods at 140 °F (60 °C) or above.

(iv) Heat treating acid or acidified foods to destroy mesophilic microorganisms when those foods are to be held in hermetically sealed containers at ambient temperatures.

(4) Measures such as sterilizing, irradiating, pasteurizing, freezing, refrigerating, controlling pH or controlling a_w that are taken to destroy or prevent the growth of undesirable microorganisms, particularly those of public health significance, shall be adequate under the conditions of manufacture, handling, and distribution to prevent food from being adulterated within the meaning of the act.

(5) Work-in-process shall be handled in a manner that protects against contamination.

(6) Effective measures shall be taken to protect finished food from contami-

nation by raw materials, other ingredients, or refuse. When raw materials, other ingredients, or refuse are unprotected, they shall not be handled simultaneously in a receiving, loading, or shipping area if that handling could result in contaminated food. Food transported by conveyor shall be protected against contamination as necessary.

(7) Equipment, containers, and utensils used to convey, hold, or store raw materials, work-in-process, rework, or food shall be constructed, handled, and maintained during manufacturing or storage in a manner that protects against contamination.

(8) Effective measures shall be taken to protect against the inclusion of metal or other extraneous material in food. Compliance with this requirement may be accomplished by using sieves, traps, magnets, electronic metal detectors, or other suitable effective means.

(9) Food, raw materials, and other ingredients that are adulterated within the meaning of the act shall be disposed of in a manner that protects against the contamination of other food. If the adulterated food is capable of being reconditioned, it shall be reconditioned using a method that has been proven to be effective or it shall be reexamined and found not to be adulterated within the meaning of the act before being incorporated into other food.

(10) Mechanical manufacturing steps such as washing, peeling, trimming, cutting, sorting and inspecting, mashing, dewatering, cooling, shredding, extruding, drying, whipping, defatting, and forming shall be performed so as to protect food against contamination. Compliance with this requirement may be accomplished by providing adequate physical protection of food from contaminants that may drip, drain, or be drawn into the food. Protection may be provided by adequate cleaning and sanitizing of all food-contact surfaces, and by using time and temperature controls at and between each manufacturing step.

(11) Heat blanching, when required in the preparation of food, should be effected by heating the food to the required temperature, holding it at this temperature for the required time, and then either rapidly cooling the food or passing it to subsequent manufacturing without delay. Thermophilic growth and contamination in blanchers should be minimized by the use of adequate operating temperatures and by periodic cleaning. Where the blanched food is washed prior to filling, water used shall be safe and of adequate sanitary quality.

(12) Batters, breading, sauces, gravies, dressings, and other similar preparations shall be treated or maintained in such a manner that they are protected against contamination. Compliance with this requirement may be accomplished by any effective means, including one or more of the following:

(i) Using ingredients free of contamination.

(ii) Employing adequate heat processes where applicable.

(iii) Using adequate time and temperature controls.

(iv) Providing adequate physical protection of components from contaminants that may drip, drain, or be drawn into them.

(v) Cooling to an adequate temperature during manufacturing.

(vi) Disposing of batters at appropriate intervals to protect against the growth of microorganisms.

(13) Filling, assembling, packaging, and other operations shall be performed in such a way that the food is protected against contamination. Compliance with this requirement may be accomplished by any effective means, including:

(i) Use of a quality control operation in which the critical control points are identified and controlled during manufacturing.

(ii) Adequate cleaning and sanitizing of all food-contact surfaces and food containers.

(iii) Using materials for food containers and food-packaging materials that are safe and suitable, as defined in § 130.3(d) of this chapter.

(iv) Providing physical protection from contamination, particularly airborne contamination.

(v) Using sanitary handling procedures.

Food and Drug Administration, HHS

(14) Food such as, but not limited to, dry mixes, nuts, intermediate moisture food, and dehydrated food, that relies on the control of a_w for preventing the growth of undesirable microorganisms shall be processed to and maintained at a safe moisture level. Compliance with this requirement may be accomplished by any effective means, including employment of one or more of the following practices:

(i) Monitoring the a_w of food.

(ii) Controlling the soluble solids-water ratio in finished food.

(iii) Protecting finished food from moisture pickup, by use of a moisture barrier or by other means, so that the a_w of the food does not increase to an unsafe level.

(15) Food such as, but not limited to, acid and acidified food, that relies principally on the control of pH for preventing the growth of undesirable microorganisms shall be monitored and maintained at a pH of 4.6 or below. Compliance with this requirement may be accomplished by any effective means, including employment of one or more of the following practices:

(i) Monitoring the pH of raw materials, food in process, and finished food.

(ii) Controlling the amount of acid or acidified food added to low-acid food.

(16) When ice is used in contact with food, it shall be made from water that is safe and of adequate sanitary quality, and shall be used only if it has been manufactured in accordance with current good manufacturing practice as outlined in this part.

(17) Food-manufacturing areas and equipment used for manufacturing human food should not be used to manufacture nonhuman food-grade animal feed or inedible products, unless there is no reasonable possibility for the contamination of the human food.

§ 110.93 Warehousing and distribution.

Storage and transportation of finished food shall be under conditions that will protect food against physical, chemical, and microbial contamination as well as against deterioration of the food and the container.

§ 110.110

Subpart F—[Reserved]

Subpart G—Defect Action Levels

§ 110.110 Natural or unavoidable defects in food for human use that present no health hazard.

(a) Some foods, even when produced under current good manufacturing practice, contain natural or unavoidable defects that at low levels are not hazardous to health. The Food and Drug Administration establishes maximum levels for these defects in foods produced under current good manufacturing practice and uses these levels in deciding whether to recommend regulatory action.

(b) Defect action levels are established for foods whenever it is necessary and feasible to do so. These levels are subject to change upon the development of new technology or the availability of new information.

(c) Compliance with defect action levels does not excuse violation of the requirement in section 402(a)(4) of the act that food not be prepared, packed, or held under unsanitary conditions or the requirements in this part that food manufacturers, distributors, and holders shall observe current good manufacturing practice. Evidence indicating that such a violation exists causes the food to be adulterated within the meaning of the act, even though the amounts of natural or unavoidable defects are lower than the currently established defect action levels. The manufacturer, distributor, and holder of food shall at all times utilize quality control operations that reduce natural or unavoidable defects to the lowest level currently feasible.

(d) The mixing of a food containing defects above the current defect action level with another lot of food is not permitted and renders the final food adulterated within the meaning of the act, regardless of the defect level of the final food.

(e) A compilation of the current defect action levels for natural or unavoidable defects in food for human use that present no health hazard may be obtained upon request from the Industry Programs Branch (HFF-326), Center for Food Safety and Applied Nutrition, Food and Drug Administration, 200 C St. SW., Washington, DC 20204.

Food and Drug Administration, HHS

§ 160.190 Frozen egg yolks.

Frozen egg yolks, frozen yolks is the food prepared by freezing egg yolks that conform to § 160.180, with such precautions that the finished food is free of viable *Salmonella* microorganisms.

PART 161—FISH AND SHELLFISH

Subpart A—General Provisions

Sec.
161.30 Declaration of quantity of contents on labels for canned oysters.

Subpart B—Requirements for Specific Standardized Fish and Shellfish

161.130 Oysters.
161.131 Extra large oysters.
161.132 Large oysters.
161.133 Medium oysters.
161.134 Small oysters.
161.135 Very small oysters.
161.136 Olympia oysters.
161.137 Large Pacific oysters.
161.138 Medium Pacific oysters.
161.139 Small Pacific oysters.
161.140 Extra small Pacific oysters.
161.145 Canned oysters.
161.170 Canned Pacific salmon.
161.173 Canned wet packed shrimp in transparent or nontransparent containers.
161.175 Frozen raw breaded shrimp.
161.176 Frozen raw lightly breaded shrimp.
161.190 Canned tuna.

AUTHORITY: Secs. 201, 401, 403, 409, 701, 706 of the Federal Food, Drug, and Cosmetic Act (21 U.S.C. 321, 341, 343, 348, 371, 376).

SOURCE: 42 FR 14464, Mar. 15, 1977, unless otherwise noted.

Subpart A—General Provisions

§ 161.30 Declaration of quantity of contents on labels for canned oysters.

(a) For many years packers of canned oysters in the Gulf area of the United States have labeled their output with a declaration of the drained weight of oysters in the containers. Packers in other areas have marketed canned oysters with a declaration of the total weight of the contents of the container. Investigation reveals that under present-day practice consumers generally do not discard the liquid packing medium, but

use it as a part of the food. Section 403(e)(2) of the Federal Food, Drug, and Cosmetic Act and the regulations thereunder require food in package form to bear an accurate label statement of the quantity of food in the container.

(b) It is concluded that compliance with the label declaration of quantity of contents requirement will be met by an accurate declaration of the total weight of the contents of the can. The requirements of § 161.145(c), establishing a standard of fill of container for canned oysters and specifying the statement of substandard fill for those canned oysters failing to meet that standard remain unaffected by this interpretation.

Subpart B—Requirements for Specific Standardized Fish and Shellfish

§ 161.130 Oysters.

(a) Oysters, raw oysters, shucked oysters, are the class of foods each of which is obtained by shucking shell oysters and preparing them in accordance with the procedure prescribed in paragraph (b) of this section. The name of each such food is the name specified in the applicable definition and standard of identity prescribed in §§ 161.131 to 161.140, inclusive.

(b) If water, or salt water containing less than 0.75 percent salt, is used in any vessel into which the oysters are shucked the combined volume of oysters and liquid when such oysters are emptied from such vessel is not less than four times the volume of such water or salt water. Any liquid accumulated with the oysters is removed. The oysters are washed, by blowing or otherwise, in water or salt water, or both. The total time that the oysters are in contact with water or salt water after leaving the shucker, including the time of washing, rinsing, and any other contact with water or salt water is not more than 30 minutes. In computing the time of contact with water or salt water, the length of time that oysters are in contact with water or salt water that is agitated by blowing or otherwise, shall be calculated at twice its actual length. Any period of time that oysters are in contact with

§ 161.131   21 CFR Ch. I (4-1-90 Edition)

salt water containing not less than 0.75 percent salt before contact with oysters, shall not be included in computing the time that the oysters are in contact with water or salt water. Before packing into the containers for shipment or other delivery for consumption the oysters are thoroughly drained and are packed without any added substance.

(c) For the purposes of this section:

(1) "Shell oysters" means live oysters of any of the species, *Ostrea virginica*, *Ostrea gigas*, *Ostrea lurida*, in the shell, which, after removal from their beds, have not been floated or otherwise held under conditions which result in the addition of water.

(2) "Thoroughly drained" means one of the following:

(i) The oysters are drained on a strainer or skimmer which has an area of not less than 300 square inches per gallon of oysters, drained, and has perforations of at least ¼ of an inch in diameter and not more than 1¼ inches apart, or perforations of equivalent areas and distribution. The oysters are distributed evenly over the draining surface of the skimmer and drained for not less than 5 minutes; or

(ii) The oysters are drained by any method other than that prescribed by paragraph (c)(2)(i) of this section whereby liquid from the oysters is removed so that when the oysters are tested within 15 minutes after packing by draining a representative gallon of oysters on a skimmer of the dimensions and in the manner described in paragraph (c)(2)(i) of this section for 2 minutes, not more than 5 percent of liquid by weight is removed by such draining.

§ 161.131   Extra large oysters.

Extra large oysters, oysters counts (or plants), extra large raw oysters, raw oysters counts (or plants), extra large shucked oysters, shucked oysters counts (or plants), are of the species *Ostrea virginica* and conform to the definition and standard of identity prescribed for oysters by § 161.130 and are of such size that 1 gallon contains not more than 160 oysters and a quart of the smallest oysters selected therefrom contains not more than 44 oysters.

§ 161.132   Large oysters.

Large oysters, oysters extra selects, large raw oysters, raw oysters extra selects, large shucked oysters, shucked oysters extra selects, are of the species *Ostrea virginica* and conform to the definition and standard of identity prescribed for oysters by § 161.130 and are of such size that 1 gallon contains more than 160 oysters but not more than 210 oysters; a quart of the smallest oysters selected therefrom contains not more than 58 oysters, and a quart of the largest oysters selected therefrom contains more than 36 oysters.

§ 161.133   Medium oysters.

Medium oysters, oysters selected, medium raw oysters, raw oysters selects, medium shucked oysters, shucked oysters selects, are of the species *Ostrea virginica* and conform to the definition and standard of identity prescribed for oysters by § 161.130 and are of such size that 1 gallon contains more than 210 oysters, but not more than 300 oysters; a quart of the smallest oysters selected therefrom contains not more than 83 oysters, and a quart of the largest oysters selected therefrom contains more than 46 oysters.

§ 161.134   Small oysters.

Small oysters, oysters standards, small raw oysters, raw oysters standards, small shucked oysters, shucked oysters standards, are of the species *Ostrea virginica* and conform to the definition and standards of identity prescribed for oysters by § 161.130 and are of such size that 1 gallon contains more than 300 oysters but not more than 500 oysters; a quart of the smallest oysters selected therefrom contains not more than 138 oysters and a quart of the largest oysters selected therefrom contains more than 68 oysters.

§ 161.135   Very small oysters.

Very small oysters, very small raw oysters, very small shucked oysters are of the species *Ostrea virginica* and conform to the definition and standard of identity prescribed for oysters by § 161.130 and are of such size that 1

Food and Drug Administration, HHS

§ 161.145

gallon contains more than 500 oysters, and a quart of the largest oysters selected therefrom contains more than 112 oysters.

§ 161.136 Olympia oysters.

Olympia oysters, raw Olympia oysters, shucked Olympia oysters, are of the species *Ostrea lurida* and conform to the definition and standard of identity prescribed for oysters in § 161.130.

§ 161.137 Large Pacific oysters.

Large Pacific oysters, large raw Pacific oysters, large shucked Pacific oysters, are of the species *Ostrea gigas* and conform to the definitions and standards of identity prescribed for oysters by § 161.130 and are of such size that 1 gallon contains not more than 64 oysters, and the largest oyster in the container is not more than twice the weight of the smallest oyster therein.

§ 161.138 Medium Pacific oysters.

Medium Pacific oysters, medium raw Pacific oysters, medium shucked Pacific oysters, are of the species *Ostrea gigas* and conform to the definition and standard of identity prescribed for oysters by § 161.130 and are of such size that 1 gallon contains more than 64 oysters and not more than 96 oysters, and the largest oyster in the container is not more than twice the weight of the smallest oyster therein.

§ 161.139 Small Pacific oysters.

Small Pacific oysters, small raw Pacific oysters, small shucked Pacific oysters, are of the species *Ostrea gigas* and conform to the definition and standard of identity prescribed for oysters by § 161.130 and are of such size that 1 gallon contains more than 96 oysters and not more than 144 oysters, and the largest oyster in the container is not more than twice the weight of the smallest oyster therein.

§ 161.140 Extra small Pacific oysters.

Extra small Pacific oysters, extra small raw Pacific oysters, extra small shucked Pacific oysters, are of the species *Ostrea gigas* and conform to the definition and standard of identity prescribed for oysters by § 161.130 and are of such size that 1 gallon contains

more than 144 oysters, and the largest oyster in the container is not more than twice the weight of the smallest oyster therein.

§ 161.145 Canned oysters.

(a) *Identity.* (1) Canned oysters is the food prepared from one or any mixture of two or all of the forms of oysters specified in paragraph (a)(2) of this section, and a packing medium of water, or the watery liquid draining from oysters before or during processing, or a mixture of such liquid and water. The food may be seasoned with salt. It is sealed in containers and so processed by heat as to prevent spoilage.

(2) The forms of oysters referred to in paragraph (a)(1) of this section are prepared from oysters which have been removed from their shells and washed and which may be steamed while in the shell or steamed or blanched or both after removal therefrom, and are as follows:

(i) Whole oysters with such broken pieces of oysters as normally occur in removing oysters from their shells, washing, and packing.

(ii) Pieces of oysters obtained by segregating pieces of oysters broken in shucking, washing, or packing whole oysters.

(iii) Cut oysters obtained by cutting whole oysters.

(3)(i) When the form of oysters specified in paragraph (a)(2)(i) of this section is used, the name of the food is "Oysters" or "Cove oysters", if of the species *Ostrea virginica;* "Oysters" or "Pacific oysters", if of the species *Ostrea gigas;* "Oysters" or "Olympia oysters", if of the species *Ostrea lurida.*

(ii) When the form of oysters specified in paragraph (a)(2)(ii) of this section is used, the name of the food is "Pieces of ———", the blank being filled in with the name "Oysters" or "Cove oysters", if of the species *Ostrea virginica;* "Oysters" or "Pacific oysters", if of the species *Ostrea gigas;* "Oysters" or "Olympia oysters", if of the species *Ostrea lurida.*

(iii) When the form of oysters specified in paragraph (a)(2)(iii) of this section is used, the name of the food is

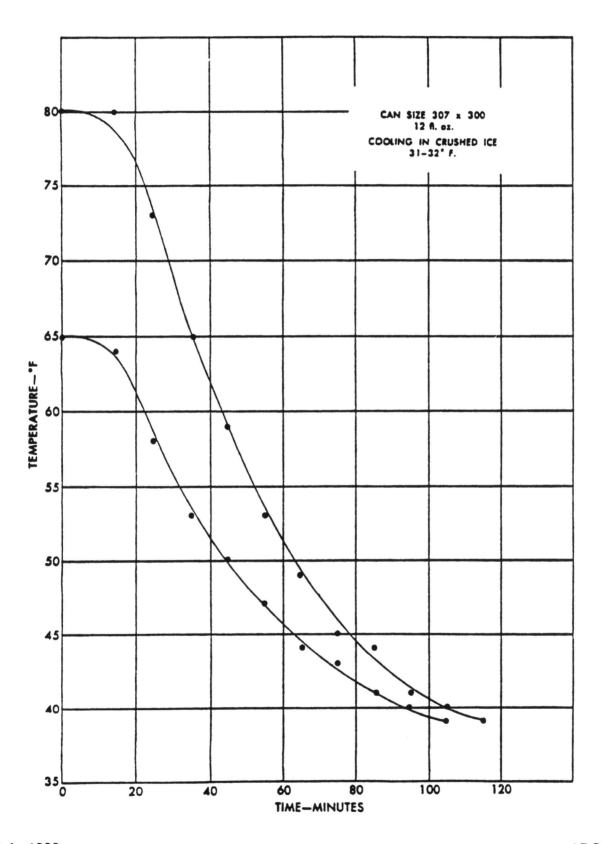

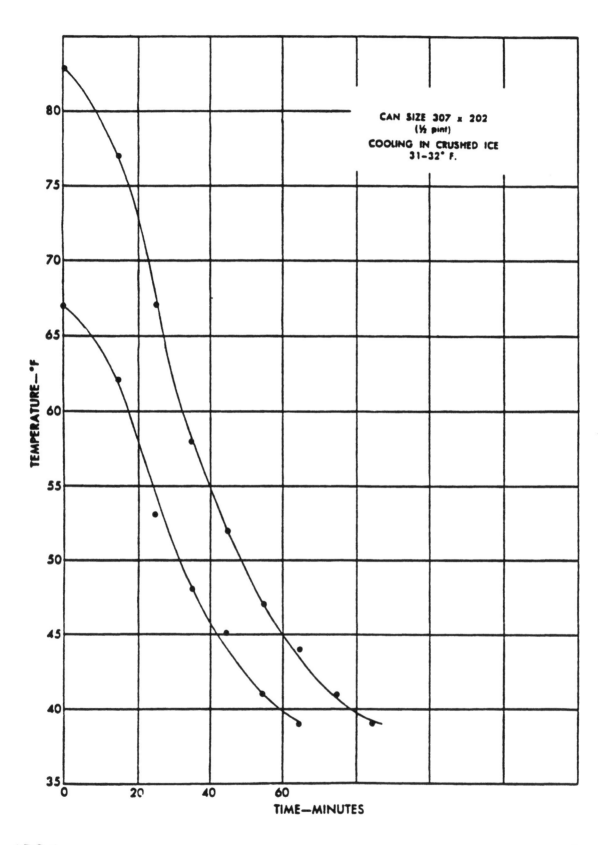

CAN SIZE 307 x 202
(½ pint)
COOLING IN CRUSHED ICE
31–32° F.

Appendix C

COOLING RATES OF FRESH OYSTERS

Central Laboratory Report[*]

Object

At the request of the USPHS the rate of cooling fresh oysters was determined on various size cans in crushed ice and under dry refrigeration.

Conclusions

The attached graphs contain the cooling rate curves for 1 gallon (610 x 708), ½ gallon (610 x 314), 1 pint (307 x 314), 12 fl. oz. (307 x 300), and ½ pint (307 x 202) cans cooled in crushed ice and cooled in a dry refrigerated chest. As expected, the cooling rate in crushed ice was faster than in dry refrigeration. Following the inital lag period, the cooling rates were generally the same regardless of initial temperatures.

Procedure

Fresh standard grade oysters were heated in a steam-jacketed kettle to the desired initial temperature and filled into the cans for the first run at each refrigeration condition. In subsequent runs the oysters were warmed in a water or air bath to the desired initial temperature.

The temperatures in the cans were taken with heat penetration thermocouples connected to a potentiometer. The junction of the thermocouple was located at the geometric center of the can.

The first cooling rate determination was made with the cans packed in crushed ice. The cans were covered with ice at all times and a drain carried away the water as the ice thawed. The ice temperature was 31°–32° F.

The second determination was made in a refrigerated chest at a temperature of 31°–32° F. A small fan in the chest kept the air gently circulating.

Fresh oysters were used for each refrigeration condition and no deterioration other than some sloughing from physical agitation was noted.

Discussion

The original request was for cooling rates at initial temperatures of 50° F. increments. We believe that from the attached curves which represent maximum and minimum initial temperatures, the time to cool to any given temperature from any given initial temperature can be interpolated very closely.

D. B. MORDEN,
Meat, Fish, and Dairy Group.

[*]Prepared by the American Can Company, Technical Services Division at the request of the U.S. Public Health Service.

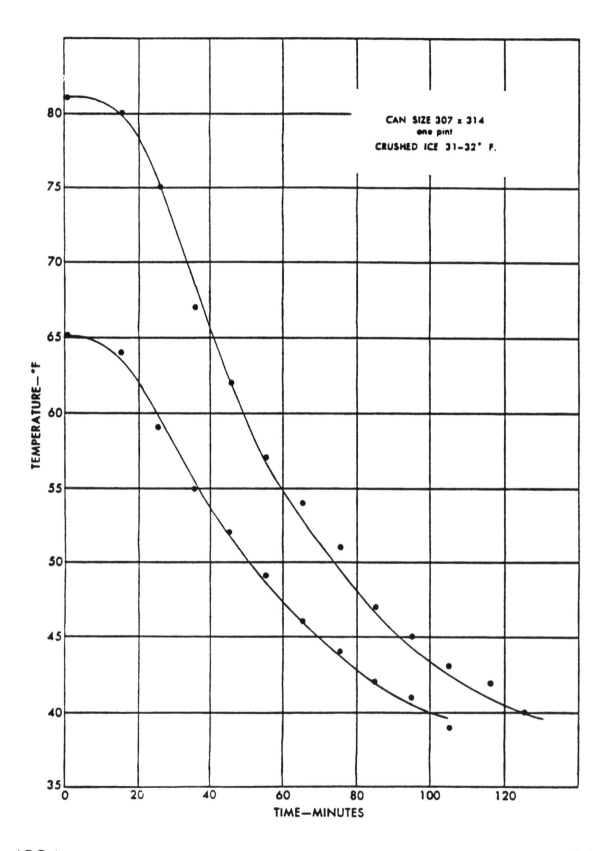

"Cut ——", the blank being filled in with the name "Oysters" or "Cove oysters", if of the species *Ostrea virginica;* "Oysters" or "Pacific oysters", if of the species *Ostrea gigas;* "Oysters" or "Olympia oysters", if of the species *Ostrea lurida.*

(iv) In case a mixture of two or all such forms of oysters is used, the name is a combination of the names specified in this paragraph (a)(3) of the forms of oysters used, arranged in order of their predominance by weight.

(b) [Reserved]

(c) *Fill of container.* (1) The standard of fill of container for canned oysters is a fill such that the drained weight of oysters taken from each container is not less than 59 percent of the water capacity of the container.

(2) Water capacity of containers is determined by the general method provided in § 130.12(a) of this chapter.

(3) Drained weight is determined by the following method: Keep the unopened canned oyster container at a temperature of not less than 68° or more than 95° Fahrenheit for at least 12 hours immediately preceding the determination. After opening, tilt the container so as to distribute its contents evenly over the meshes of a circular sieve which has been previously weighed. The diameter of the sieve is 8 inches if the quantity of the contents of the container is less than 3 pounds, and 12 inches if such quantity is 3 pounds or more. The bottom of the sieve is woven-wire cloth that complies with the specifications for such cloth set forth under "2.38 mm (No. 8)" in "Official Methods of Analysis of the Association of Official Analytical Chemists," 13th Ed. (1980), Table 1, "Nominal Dimensions of Standard Test Sieves (U.S.A. Standard Series)," under the heading "Definitions of Terms and Explanatory Notes," which is incorporated by reference. Copies may be obtained from the Association of Official Analytical Chemists, 2200 Wilson Blvd., Suite 400, Arlington, VA 22201-3301, or may be examined at the Office of the Federal Register, 1100 L St. NW., Washington, DC 20408. Without shifting the material on the sieve, so incline the sieve as to facilitate drainage. Two minutes from the time

drainage begins, weigh the sieve and the drained oysters. The weight so found, less the weight of the sieve, shall be considered to be the drained weight of the oysters.

(4) If canned oysters fall below the standard of fill of container prescribed in paragraph (a) of this section, the label shall bear the general statement of substandard fill specified in § 130.14(b) of this chapter in the manner and form therein specified, followed by the statement, "A can of this size should contain —— oz. of oysters. This can contains only —— oz.", the blanks being filled in with the applicable figures.

[42 FR 14464, Mar. 15, 1977, as amended at 47 FR 11832, Mar. 19, 1982; 49 FR 10102, Mar. 19, 1984; 54 FR 24895, June 12, 1989]

(This page is blank.)

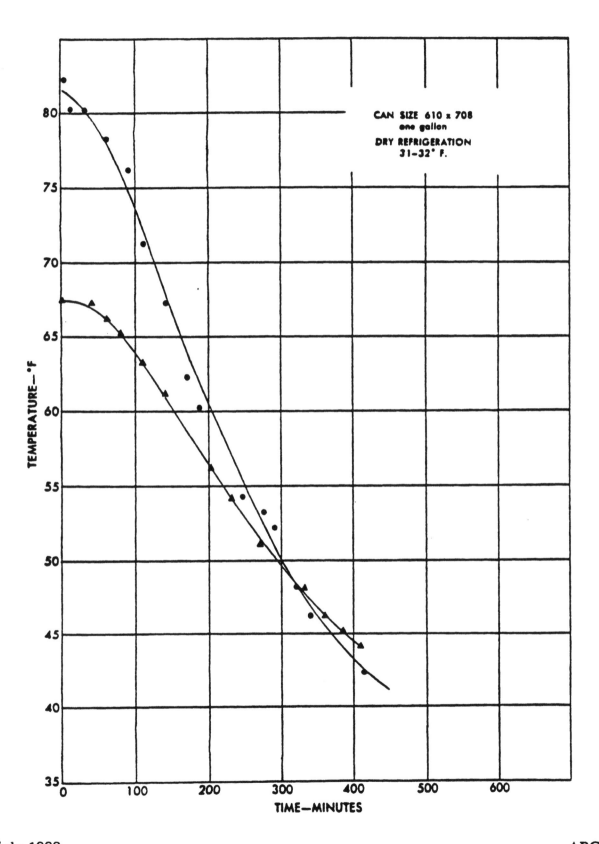

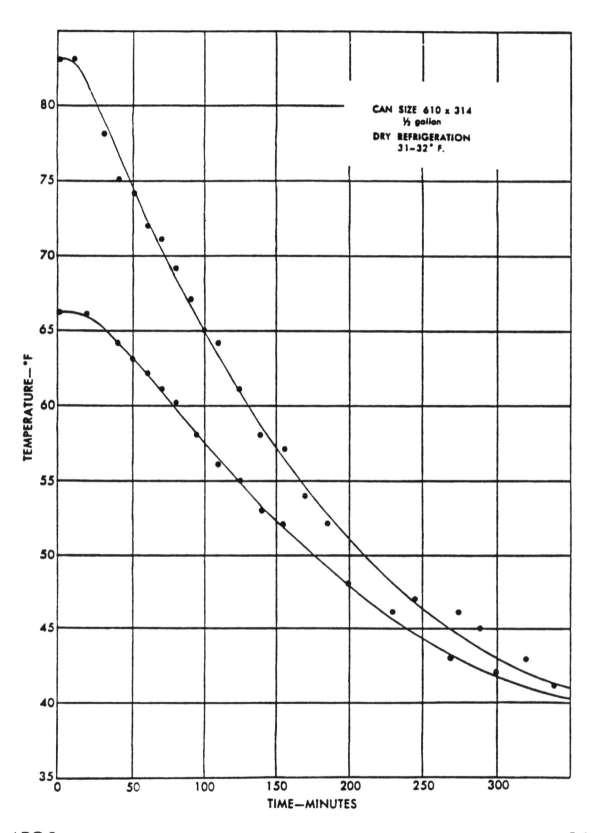

CAN SIZE 610 x 314
½ gallon
DRY REFRIGERATION
31-32° F.

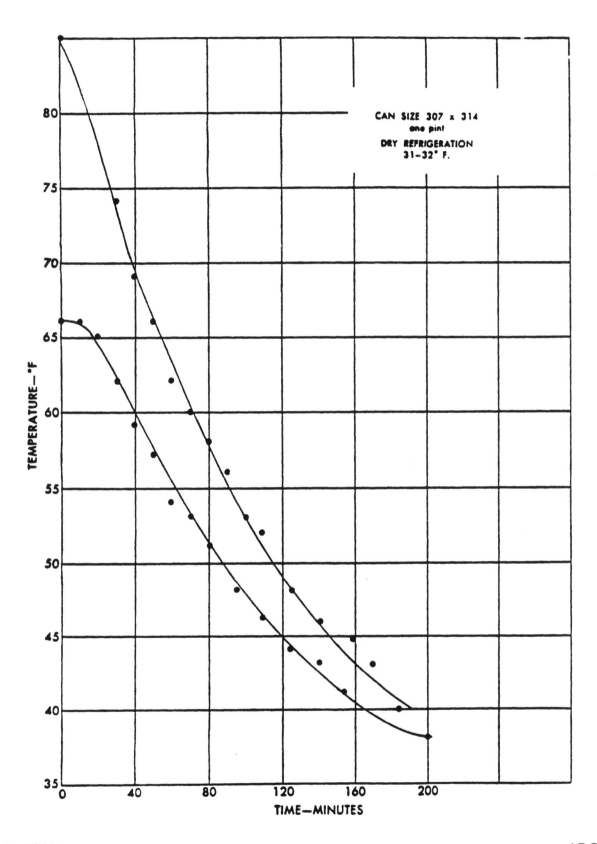

CAN SIZE 307 x 314
one pint
DRY REFRIGERATION
31–32° F.

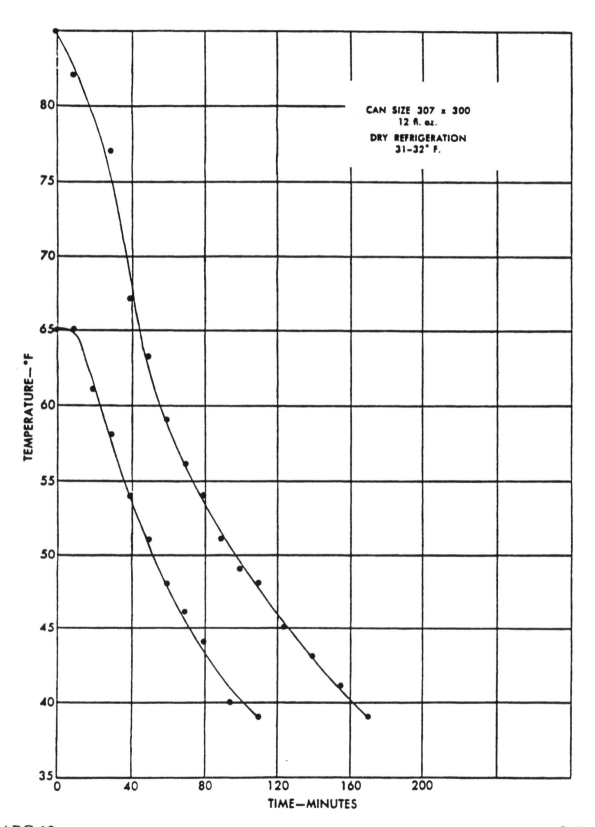

CAN SIZE 307 x 300
12 fl. oz.

DRY REFRIGERATION
31–32° F.

TEMPERATURE—°F

TIME—MINUTES

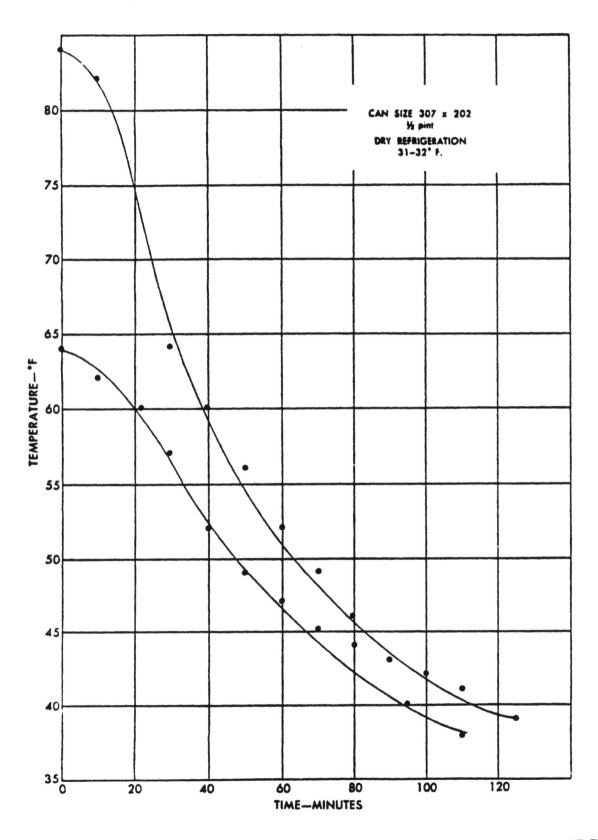

CAN SIZE 307 x 202
½ pint
DRY REFRIGERATION
31-32° F.

(This page is blank.)

Appendix D

Shellfish Ledger Records

COMPANY NAME:_____

CERTIFICATE NO.:_____

SHELLFISH HARVEST RECORD

HARVEST AREA	HARVEST DATE	SPECIES	QUANTITY

Page_____

(This page is blank.)

SHELLFISH HARVEST / PURCHASE RECORD

CERT NO.	QUANTITY	SPECIES	HARVEST AREA	HARVEST DATE	PURCHASE DATE	HARVESTER CERT #

Page _____

(This page is blank.)

SHELLFISH SALES RECORD

Page___

DATE SOLD	SOLD TO	QUANTITY	SPECIES	HARVEST DATE	HARVEST AREA	HARVESTER CERT. #

(This page is blank.)

INSPECTION FORM		
Sections B - Harvesting and Handling Shellstock, C - Wet Storage, D - Shucking and Packing Shellfish, E - Shellstock Shipping, F - Repacking, G - Reshipping, H - Heat Shock	MANUAL REFERENCE	*

			Manual Reference	Code
PLANT & GROUNDS	1.	Plant not operating under flood conditions	D, E, F, G	C
	2.	Processing operations separated by partitions, space or time	D, E, F, G	K
	3.	Storage facilities for employees, used	D, E, F, G	O
	4.	Plant premises: clean, refuse containers, stored equipment, litter, drainage	D, E, F, G	O
PLANT INTERIOR	5.	Floors: impervious, adequate drainage, maintained, clean	B, C, D, E, F, G	O
	6.	Walls, ceilings, attached equip.: smooth, light colored, clean, good repair	C, D, E, F, G	O
VECTORS	7.	Insects, rodents, vermin, other animals: excluded, controlled	B, C, D, E, F, G	K
UTILITIES	8.	Lighting adequate, fixtures shielded	C, D, E, F, G	O
	9.	Heating, cooling, ventilation adequate	D, E, F, G	O
WATER	10.	Water supply: safe source, protected, no cross connections	B, C, D, E, F, G, H	C
	11.	Adequate quantity, temperature, and pressure of water	C, D, E, F, G, H	O
PLUMBING	12.	Plumbing: meets code, adequate, functional, maintained	B, C, D, E, F, G	K
	13.	Protection against backflow, backsiphonage	C, D, E, F, G	K
	14.	Toilets: location, repair, clean, adequate number, self-closing doors, paper	D, E, F, G	K
	15.	Handwashing: location, repair, clean, soap, sanitary towels, waste receptacles, handwashing signs posted	D, E, F, G	O
SEWAGE	16.	Sewage disposal system: installed, maintained, meets code, adequate	B, C, D, E, F, G	C
CHEMICALS	17.	Poisonous/toxic materials: properly used, stored, separated, labeled	D, E, F, G	K
EQUIPMENT & UTENSILS	18.	Food contact surfaces: properly constructed and located, clean, maintained, protected from contamination	B, C, D, F, H	K
	19.	Non food contact surfaces: properly constructed and located, maintained, clean	B, D, E, F, G, H	O
	20.	Blower air intake, approved filter	D, F	O
	21.	Refrigeration units adequate, temperature measuring devices	B, D, E, G	K
CLEANING & SANITIZING	22.	Facilities: detergents, brushes, three compartment sinks, test kit, approved sanitizers	B, C, D, E, F, G, H	O
	23.	Food contact surfaces cleaned and sanitized, effective	B, D, F, H	K
SHELLFISH HANDLING & STORAGE	24.	Shellfish from approved source	C, D, E, F, G	C
	25.	Shellstock properly identified	B, C, D, E, F, G	K
	26.	Shellstock: wholesome, proper temperature	C, D, E, F	C
	27.	Shellstock clean	B, C, D, E, G, H	K
	28.	Shellstock: protected from contamination and deterioration, separated by lot, no commingling	B, C, D, E, G, H	K
	29.	Wet storage: approved	C	C
	30.	Shellfish not contaminated	B, C, D, E, F, G, H	C
	31.	Dip buckets not used	D, F	O
	32.	Single service containers: clean, stored	D, F, G	O
	33.	Containers properly labeled	D, F, G	K
	34.	Returnable containers properly used	D, F	O
	35.	Shellfish promptly shucked, packed and protected from contamination	D, F, G	K
	36.	Shucked shellfish cooled to 45 F (7.2 C) within time limits	D, F, H	C
	37.	Shucked shellfish maintained at 45 F (7.2 C) or less during storage and repack	D, F, G, H	C
	38.	Frozen shellfish maintained at 0 F (-17.8 C) or less	D, G	O
	39.	Ice: approved source, sanitary, properly protected	D, E, F, G	C
PERSONNEL	40.	Hands washed/sanitized, good hygienic practices	D, E, F, G	K
	41.	Clean outer garments. Gloves, finger cots, other coverings impermeable, sanitized as necessary and properly stored. Hair restraints	D, F	O
	42.	Personnel with infections restricted	D, F	K
	43.	Unauthorized persons prohibited	D, F	O
WASTE	44.	Shell waste: promptly removed	D, F	O
	45.	Water disposal: meets code, adequate, installed	D, E, F, G	O
SUPERVISION	46.	Supervision: responsible person designated, effective	C, D, E, F, G, H	K
RECORDS	47.	Records: complete and maintained	C, D, E, F, G	K

* CODES: (C) Critical Item   (K) Key Item   (O) Other

INSPECTION REPORT - SHELLFISH PROCESSING PLANT

Pre-Operational [] · Routine []   Follow-Up []   Standardization []

Date: Time Begin: Time End:

Firm Name: Certif. No:

Plant Location:

Plant Rep. Name: Title:

Inspector Signature: Agency:

Failure to comply with time limits for corrections of deficiencies specified in this report or through subsequent notification may result in cessation of your operation and withdrawal of certification as described in the NSSP Manual Part II, A.2.e and f.

ITEM NO.	REMARKS	CORRECTION DATE

References

1. Frost, W. H., Chairman. Report of Committee on Sanitary Control of the Shellfish Industry in the United States, Supplement No. 53, Public Health Reports, Nov. 6, 1925. 17 pp. Available from: FDA Northeast Technical Services Unit, Bldg. S-26, North Kingstown, RI 02852.

2. Jensen, E. T. The 1954 National Conference on Shellfish Sanitation, [first national workshop]. Public Health Reports, 70(9); 1955.

3. Jensen, E. T., ed. Proceedings - 1956 Shellfish Sanitation Workshop, (second national workshop), August 27-28; Washington, D.C. 143 p. Available from: FDA, Northeast Technical Services Unit, Bldg. S-26, North Kingstown, RI 02852.

4. Jensen, E. T., ed. Proceedings - 1958 Shellfish Sanitation Workshop (third national workshop) August 26-27; Washington, D.C., 72 p. Available from: FDA, Northeast Technical Services Unit, Bldg. S-26, North Kingstown, RI 02852.

5. Jensen, E. T., ed. Proceedings - 1961 Shellfish Sanitation Workshop, (fourth national workshop) November 28-30; Washington, D.C. 288 p. Available from: National Technical Information Services, Dept. of Commerce, 5285 Port Royal Rd. Springfield, VA 22161. PB8 6 236875/AS.

6. Houser, L. S., ed. Proceedings - Fifth National Shellfish Sanitation Workshop, 1964, November 17-19; Washington, D.C. 239 p. Available from: National Technical Information Services, Dept. of Commerce, 5285 Port Royal Rd. Springfield, VA 22161. PB8 6 236882/AS.

7. Morrison, G., ed. Proceedings - Sixth National Shellfish Sanitation Workshop, 1968, February 7-9; Washington, D.C., 115 p. Available from: National Technical Information Services, Dept. of Commerce, 5285 Port Royal Rd. Springfield, VA 22161. PB8 6 236890/AS.

8. Ratcliffe, S. D. and Wilt, D. S., eds. Proceedings - Seventh National Shellfish Sanitation Workshop, 1971, Oct. 20-22, Washington, D.C., 412 p. Available from: National Technical Information Services, Dept. of Commerce, 5285 Port Royal Rd. Springfield, VA 22161. PB8 6 236908/AS.

9. Wilt, D. S., ed. Proceedings - Eighth National Shellfish Sanitation Workshop, 1974, Jan. 16-18; New Orleans, LA, 158 p. Available from: National Technical Information Services, Dept. of Commerce, 5285 Port Royal Rd. Springfield, VA 22161. PB8 6 236916/AS.

10. Wilt, D. S., ed. Proceedings - Ninth National Shellfish Sanitation Workshop; 1975, June 25-26; Charleston, SC, 150 p. Available from: FDA, Northeast Technical Services Unit, Bldg. S-26, North Kingstown, RI 02852.

11.   Wilt, D. S., ed. Proceedings - Tenth Shellfish Sanitation Workshop; 1977, June 29-30, Hunt Valley, MD   236 p.   Available from: FDA, Northeast Technical Services Unit, Bldg. S-26, North Kingstown, RI   02852.

12.   Food and Drug Administration, Proposed National Shellfish Safety Program regulations. Federal Register, Vol. 40, No. 119, Thurs., June 19, 1975.

13.   Presnell, M. W. and Kelly, C. B.   Bacteriological Studies of Commercial Shellfish Operations on the Gulf Coast. U.S. Public Health Service Technical Report F-61-9; 1961.

14.   The Influence of Time and Temperature on the Bacteriological Quality of Shell Oysters During Processing and Shipping. U.S. Public Health Service, Food and Drug Administration Gulf Coast Technical Services Unit, Dauphin Island, AL; 1971. Available from: FDA Shellfish Sanitation Branch, 200 'C' Street, S.W., Washington, D.C.   20204.

15.   Bacteriological Quality of Approved Area Summer Harvested Louisiana Oysters During Harvest and Interstate Shipment. U.S. Department of Health and Human Services, Public Health Service, Food and Drug Administration, Shellfish Sanitation Branch, Northeast Technical Services Unit, North Kingstown, RI; 1983.   85 p. Available from: FDA, Shellfish Sanitation Branch, 200 'C' Street, S.W., Washington, D.C.   20204.

16.   Musselman, J. Bacteriological Study of Water and Soft Shell Clam, Mya arenaria in the Miles River, Maryland. U.S. Department of Health, Education and Welfare, Food and Drug Administration, Shellfish Sanitation Branch, Northeast Technical Services Unit, North Kingstown, RI; 1983.   Available from: FDA, Northeast Technical Services Unit, Bldg. S-26, North Kingstown, RI   02852.

17.   Kelly, C. B. and Arcisz, W.   Bacteriological Control of Oysters During Processing and Marketing. Public Health Reports, Vol. 69, No. 8; 1954.

18.   Kelly, C. B. and Arcisz, W.   Survival of Enteric Organisms in Shellfish.   Public Health Report, 69:1205-1210; 1954.

19.   Food and Drug Administration.   Special Report.   The Influence of Time and Temperature on the Bacterial Quality of Shell Oysters During Processing and Shipping, 1971. U.S. DHEW Public Health Service, Food and Drug Administration, Gulf Coast Technical Services Unit, 15 p; 1971. Available from: FDA, Shellfish Sanitation Branch, 200 'C' Street, S.W., Washington, D.C.   20204.

20.   Clem, J. D. Status of the Recommended National Shellfish Sanitation Program Bacteriological Criteria for Shucked Oysters at the Wholesale Market Level, DHHS, PHS, FDA, BF, SSB, Washington, D.C. 1982. Available from: FDA, Shellfish Sanitation Branch, 200 'C' Street, S.W., Washington, D.C.   20204.

21.   Hunter, A.C. and C.W. Harrison. Bacteriology and Chemistry of Oysters, with Special Reference to Regulatory Control of Production, Handling, and Shipment. United States Department of Agriculture, Washington, D.C. Technical Bulletin No. 64, March, 1928.

22.   Federal Security Agency, United States Public Health Service, Manual of Recommended Practice for Sanitary Control of the Shellfish Industry, Public Health Bulletin No. 295, U.S. Government Printing Office, Washington, D.C., 1946.

23.   U.S. Department of Health, Education and Welfare (1965), Public Health Service, National Shellfish Sanitation Program Manual of Operations, Part I - Sanitation of Shellfish Growing Areas 1965 Revision, Public Health Service Publication No. 33.

24.   Cook, D.W. and Ruple, A.D. Indicator Bacteria and Vibrionaceae Multiplication in Post-Harvest Shellstock Oysters. Jour. Food Prot. 52:343-349; 1989.

25.   Standard Methods for the Examination of Water and Wastewater. 17th ed., Washington, D.C. American Public Health Association, American Water Works Association, Water Pollution Control Federation; 1989.

26.   Williams, L. A. and LaRock, P. A. Temporal Occurence of Vibrio species and Aeromonas hydrophila In Estuarine Sediments. Appl. Environ. Microbiol. 50:1490-1495; 1985.

27.   Liston, J.   Influence of U.S. Seafood-Handling Procedures On Vibrio parahaemolyticus. International Symposium on Vibrio parahaemolyticus. Eds. T. Fujino, G. Sakaguchi, R. Sakazaki, and Y. Takeda. Tokyo, Saikon: 123-128; 1974.

28.   Boutin, B. K., et al. Effect of Temperature and Suspending Vehicle on Survival of Vibrio parahaemolyticus and V. vulnificus. Jour. Food Prot. 48:875-878; 1985.

29.   Vaughn, M. W. and Jones, W. W. Bacteriological Survey of an Oyster Bed in Tangier Sound, Maryland. Chesapeake Science, (5) 4; 1964.

30.   Roos, B. Hepatitis Epidemic Conveyed by Oysters. Svenska Lakartodningen 53 (16):989-1003; 1956.   (Translation available from the Food and Drug Administration, Shellfish Sanitation Branch (HFF-344), Washington, D.C. 20204.

31.   Food and Drug Administration.   England, Shellfish Program Review - 1983. (Technical Report) 1984; p. 68. Available from: FDA, Northeast Technical Services Unit, Bldg. S-26, North Kingstown, RI 02852.

32.   Son, N. T. and Fleet, G. H. Behavior of Pathogenic Bacteria in the Oyster, Crassostrea commercialis, During Depuration, Relaying and Storage. App. and Environ. Micro. 40 (6):994-1022; 1980.

33.   Nolan, Charles M., et al. _Vibrio parahaemolyticus_ Gastroenteritis - An Outbreak Associated With Raw Oysters In the Pacific Northwest.   Diagnosis of Microbiological Infectious Diseases, 2:119-128; 1984.

34.   Wilson, R., et al. Non-01 group 1 _Vibrio cholerae_ Gastroenteritis Associated With Eating Raw Oysters. Am J Epidemiol. 114:293-8; 1981.

35.   Boutin, B. K., et al. Effect of Temperature and Suspending Vehicle on Survival of _Vibrio parahaemolyticus_ and _V. vulnificus_. Jour. Food Prot. 48:875-878; 1985.

36.   Code of Federal Regulations, Title 21, Parts 170-199.   Superintendent of Documents, U.S. Government Printing Office, Washington, D.C. 20402; 1986.

37.   Rindge, M. E., et al. A Case Study on the Transmission of Infectious Hepatitis by Raw Clams. U.S. Department of Health, Education and Welfare, Public Health Service, Washington, D.C.; 1962.

38.   Murphy, A. M., et al.   An Australia-Wide Outbreak of Gasteroenteritis from Oysters Caused by Norwalk Virus. Med. J. Aust. 2:329-333; 1979.

39.   Casper, V. Memorandum, Quarterly Report - Shellfish, October 1, 1982, through December 31, 1982, Oyster Related Gastroenteritis Outbreaks. DHHS, Food and Drug Administration, Dallas, TX; 1982. Available from: FDA, Shellfish Sanitation Branch, 200 'C' Street, S.W., Washington, D.C. 20204.

40.   Old, H. N. and Gill, S. L. A Typhoid Fever Epidemic Caused by Carrier Bootlegging Oysters. Am. J. of Public Health, 30:633-640; 1940.

41.   Galtsoff, P. S. Biology of the Oyster in Relation to Sanitation. Am. J. of Public Health. 26:245-247; 1936.

42.   Cabelli, V. J. and Heffernan, W. P. Seasonal Factors Relevant to Coliform Levels in the Northern Quahog. Proceedings of the National Shellfisheries Association, 61:95-101; 1971.

43.   Fisher, L. M. and Acker, J. E. Bacteriological Examination of Oysters and Water from Narragansett Bay During the Winter and Spring in 1927-28 Public Health Reports. 50 (42); 1935.

44.   Pringle, B., et al. Trace Metal Accumulation by Estuarine Mollusks. J. San. Eng. Div. ASCE. 94:455-475; 1968.

45.   Studies of the Fate of Certain Radionuclides in Estuarine and Other Aquatic Environments. Public Health Service Publication No. 999-R-3. Washington, D.C., 73 p; 1963. Available from: FDA, Shellfish Sanitation Branch, 200 'C' Street, S.W., Washington, D.C. 20204.

46.   Gordon, K., M.D., et al. Shellfish Poisoning. Center for Disease Control, Morbidity and Mortality Weekly Report, 22, (48):397-398; 1973.

47.   Eldred, B., et al. Preliminary Studies of the Relation of **Gymnodinum Breve** Counts to Shellfish Toxicity. A Collection of Data in Reference to Red Tide Outbreaks During 1963, Reproduced by the Marine Laboratory of the Florida Board of Conservation, St. Petersburg, FL, May 1964.

48.   Kopfler, F. C. and Mayer, J. Studies on Trace Metals in Shellfish. Proceedings, Gulf and South Atlantic Research Conference, March 1967, Gulf Coast Marine Health Science Laboratory, Dauphin Island, AL; 1969. Available from: FDA, Northeast Technical Services Unit, Bldg. S-26, North Kingstown, RI 02852.

49.   Galtsoff, P. S. The American Oyster. Fishery Bulletin No. 64. U.S. Department of the Interior, U.S. Government Printing Office, Washington, D.C. 20402; 1964.

50.   Fraiser, M. B. and Koburger, J. A. Incidence of Salmonellae in Clams, Oysters, Crabs and Mullet. Journal of Food Protection(47) 5:343-345; 1984.

51.   Kutsuna, Chief of study group., Minamata Disease, Kumamoto, Japan. Kumamoto University; 1968.

52.   Thrower, S. J. and Eustace, I. J. Heavy Metal Accumulation in Oysters Grown in Tasmanian Waters. Food Technology in Australia 25 (11): 546-553; 1973.

53.   Medcof, J. C., et al. Paralytic Shellfish Poisoning on the Canadian Atlantic Coast. Bulletin of the Fisheries Research Board of Canada, 75: 32 p; 1947.

54.   Prakash, A., et al. Paralytic Shellfish Poisoning in Eastern Canada. Fisheries Research Board of Canada, Bulletin No. 177, 87 p; 1971.

55.   Schwalm, D. J. The 1972 PSP Outbreak in New England. Report, DHEW, Food and Drug Administration, Boston, MA; 1973.

56.   LoCicero, V. R., ed. Proceedings of the First International Conference on Toxic Dinoflagellate Blooms. Massachusetts Science and Technology Foundation, Wakefield, MA; 1975.

57.   Koff, R. S. and Sear, H. S. Internal Temperature of Steamed Clams. New England J. Med. 276:737-739; 1967.

58.   Tennant, A. D. An Investigation of the Bacterial Flora of the Soft Shell Clam (Mya arenaria) in New Brunswick and Nova Scotia. Montreal, Canada, McGill University, 1949. Master of Science Thesis. Abstract available from: Shellfish Sanitation Branch, FDA, 200 'C' Street, S.W., Washington, D.C. 20204.

59. Cook, D. W., et al. Steam Unit to Aid in Oyster Shucking. Proc. Fifth Ann. Trop. Subtrop. Fish. Tech. Conf., Sept. 1980.

60. Food and Drug Administration. Field and Laboratory Studies on Heat-Shock Method of Preparation of Oysters for Shucking. U.S. Dept. HEW, PHS, Gulf Coast Shellfish Sanitation Research Center, Dauphin Island, AL; 1964. 27p. Available from: Food and Drug Administration, Shellfish Sanitation Branch, 200 'C' Street, S.W., Washington, D.C. 20204.

61. Fleet, G. H. Oyster Depuration - A Review. Food Technol. Aust. 30:444; 1978.

62. Furfari, S. A. Depuration Plant Design. Public Health Service Publication. (999-FP-7):119p; 1966.

63. Heffernan, P. and Cabelli, V. The Elimination of Bacteria by the Northern Quahog: Variability in the Response of Individual Animals and the Development of Criteria. Proc. Nat. Shell. Assoc. 61:102-108; 1971.

64. Ayres, P. A.; Shellfish Purification in Installations Using Ultraviolet Light. Laboratory Leaflet No. 43, U. K. Ministry of Agriculture, Fisheries and Food, England, 1978.

65. Haven, D. S., et al. Bacterial Depuration by the American Oyster (Crassostrea virginica) Under Controlled Conditions, Volume I, Biological and Technical Studies, Virginia Institute of Marine Science, Gloucester Point, VA, Special Scientific Report No. 88:62 p; May 1978. Available from: NTIS Accession No. PB86-172194/AS.

66. Heffernan, P. and Cabelli, V. The Elimination of Bacteria by the Northern Quahog (Mercenaria mercenaria): Environmental Parameters Significant to the Process. J. Fish Res. Rd. Canada. 27:1569-1577i; 1970.

67. Metcalf, T. G. et al. Bioaccumulation and Depuration of Enteroviruses by the Soft-Shelled Clam, Mya arenaria. App. and Envir. Micro. 38 (2): 275-282; 1979.

68. Rowse, A. J. and Fleet, G. N. Effects of Water Temperature and Salinity on Elimination of Salmonella charity and Escherichia coli from Sydney Rock Oysters (Crassostrea commercialis). Appl. Environ. Micro. 48 (5):1061-1063; 1984.

69. Furfari, S. A. Shellfish Purification: A Review of Current Technology, FAO Technical Conference on Aquaculture, FIR: AQ/Conf/76R. 11. 16 pages; 1976.

70. Kelly, C. B. Disinfection of Sea Water by Ultraviolet Radiation. Am. J. Public Health, 51(11):16-70-1680; 1961.

71. Furfari, S. A. Appendix D - Controlled Purification. Proc. Seventh National Workshop, p. 333-347; 1971.

72. Canzonier, W. J. Accumulation and Elimination of Coliphage S-13 by the Hard Clam, <u>Mercenaria mercenaria</u>. App. Micro. 21 (6):1024-1031; 1971.

73. Summary of Actions contained in letter from Chairman, ISSC to Chief, Shellfish Sanitation Branch, FDA, November 14, 1984.

74. Lake, R.L. FDA Office of Compliance, [Letter to Charles Conrad, Louisiana Dept. of Health and Hospitals] Subject: Use of Ozone or 'Photozone' in Wet Storage and Depuration. August 18, 1989. Available from: FDA, Shellfish Sanitation Branch, 200 C St., S.W. Washington, D.C. 20204.

75. Summary of Actions contained in letter from Chairman, ISSC to Chief, Shellfish Sanitation Branch, FDA, November 4, 1985.

☆ U.S. GOVERNMENT PRINTING OFFICE:1992-617-023/68012

Milton Keynes UK
Ingram Content Group UK Ltd.
UKHW020829141024
449569UK00030B/1539